The Human Brain during the First Trimester 3.5- to 4.5-mm Crown-Rump Lengths

This first of 15 short atlases reimagines the classic 5-volume *Atlas of Human Central Nervous System Development*. This volume presents serial sections from specimens between 3.5 mm and 4.5 mm with detailed annotations, together with 3D reconstructions. An introduction summarizes human CNS development by using high-resolution photos of methacrylate-embedded rat embryos at a similar stage of development as the human specimens in this volume. The accompanying Glossary gives definitions for all the terms used in this volume and all the others in the *Atlas*.

Key Features

- Classic anatomical atlases
- Detailed labeling of structures in the developing brain offers updated terminology and the identification of unique developmental features, such as, germinal matrices of specific neuronal populations and migratory streams of young neurons
- Appeals to neuroanatomists, developmental biologists, and clinical practitioners
- A valuable reference work on brain development that will be relevant for decades

ATLAS OF
HUMAN CENTRAL NERVOUS SYSTEM DEVELOPMENT
Series

The Human Brain during the First Trimester 3.5- to 4.5-mm Crown-Rump Lengths

Atlas of Human Central Nervous System Development, Volume 1

Shirley A. Bayer and Joseph Altman

CRC Press
Taylor & Francis Group
Boca Raton London New York

CRC Press is an imprint of the
Taylor & Francis Group, an **informa** business

First edition published 2023
by CRC Press
6000 Broken Sound Parkway NW, Suite 300, Boca Raton, FL 33487-2742

and by CRC Press
4 Park Square, Milton Park, Abingdon, Oxon, OX14 4RN

CRC Press is an imprint of Taylor & Francis Group, LLC

LCCN no. 2022008216

ISBN: 978-1-032-18326-8 (hbk)
ISBN: 978-1-032-18325-1 (pbk)
ISBN: 978-1-003-27060-7 (ebk)

DOI: 10.1201/9781003270607

Typeset in Times Roman by KnowledgeWorks Global Ltd.

Access the support material at: https://routledge.com/9781032183268

CONTENTS

PREFACE

The first edition of the 5-volume *Atlas of Human Central Nervous System Development* began in the late 1980s with a phone call from Serge Duckett to my laboratory in Indianapolis. He asked me to write a chapter on human brain development in a book he was planning to publish (*Pediatric Neuropathology*, 1995, Williams and Wilkins, Baltimore). I began to explore the Indiana University Medical Library for some research articles or books on human brain development. To my surprise, there was very little information. Luckily, one of the lab instructors in the human anatomy course I taught was an assistant to Dr. William DeMyer, the late pediatric neurologist on the IU Medical School faculty. He kindly lent me some very precious books from his personal library which contained drawings and photos of some developing human brains. Among them was a fragile copy of Hochstetter's 1919 book: *Beitrage zur Entwicklungsgeschichte des menschlichen Gehirns*, Franz Deuticke, Wein. At that time, I had already been studying embryonic and fetal rat brains for years, working with my husband, J. Altman. Hochstetter's book was a goldmine with several reconstructions, drawings, and photos of serially-sectioned human embryonic brains. I realized that embryonic human brains and rat brains had similar morphology, and it was easy to identify structures. Both Joe and I became engrossed in the possibility that our research on rat brains could be directly applicable to human brains. The chapter I contributed to Serge's book used morphological matching to extrapolate the experimental data on rat brain development to descriptive data on human brain development (Chapter 5, Embryology, pp. 54–107). But all this was done without my having examined human specimens directly.

In 1995, I had an opportunity to work at the National Institutes of Health in Bethesda, MD, assisting a researcher in mouse brain development to publish an atlas on chemical markers. I agreed to participate if I was also given time to explore the Collections of human embryos and fetuses at the National Museum of Health and Medicine, then housed at the Armed Forces Institute of Pathology in Walter Reed Hospital, Washington DC. I spent that year taking photos of specimens in the Carnegie, Minot, and Yakovlev Collections. In the summer of 1996, Joe and I rented an apartment in DC and went every weekday to take more pictures of specimens. We took over 10,000 photos that became the database for the 1st edition of the *Atlas*, published by CRC Press between 2002 and 2008.

The 1st edition was packaged into 5 volumes: 1 (2002) on the entire span of spinal cord development in embryonic, fetal, and early postnatal age groups. We started the brain studies with the most mature specimens in the third trimester because we could more accurately identify structures (Volume 2, 2004). Then, we worked our way through progressively more immature specimens in the second trimester (Volume 3, 2005), the late first trimester (Volume 4, 2006), and ending with the most immature specimens in the early first trimester (Volume 5, 2008). While the 1st edition was being published, and certainly by the time it was completed, the publishing industry was changing drastically from exclusively paper books to more and more electronic books. For several years, the 5-volume set of the *Atlas* has been out of print. That worried me because I felt that all the work our laboratory spent on this much-needed analysis was going to be lost.

The road to the 2nd edition of the *Atlas* was long and circuitous. As soon as we finished the 1st edition of the *Atlas*, I started to put our archival work on rat brain development into websites so that students could find it. The website neurondevelopment.org contained our scientific papers in portable document format for download. The website braindevelopmentmaps.org contained a histological database of the rat brains processed in our laboratory. One set was our ^{3}H-thymidine autoradiographic series. Another set was methacrylate-embedded embryos that contained excellent histology akin to low-magnification electron microscopy. It took so long to finally complete the methacrylate set that most of our papers were based on the older paraffin-embedded developmental series. I found a laboratory in Orlando that scanned entire sections in a grid pattern using a microscope with a 40x objective attached to a computer that knitted the fields together. An observer could view sections from very low magnification to high magnification and see individual cells and fibers. We felt compelled to share this work with everybody and paid approximately $24,000 to have specimens scanned from embryonic days 10 and up to the time of birth. These scanned images are available on braindevelopmentmaps.org. The website brainmindevolution.org contained Joseph Altman's book on brain and mind evolution that he had been working on for 70 years, starting before he left Hungary after World War II and still working on it at the time of his death in 2016. I finished the last chapter (13) and the voluminous bibliography by the end of 2016.

After Joe's death, I was devastated and couldn't do any scientific work for about three to four years. Then I got an email from Dr. Jo Bhattacharya, a pediatric neurosurgeon from Glasgow, Scotland, asking me for more information about surrounding tissues in human embryos and fetuses. I went back to those 10,000 negatives and started my scientific work again! This time, I wanted to share the human development photos with everybody on a new website, brainimages.org. Still upset that the Atlas was out of print and very hard to get, I wrote to CRC Press/Taylor and Francis for permission to use the annotated images in brainimages.org. The editor, Chuck Crumly, wanted to know more about what had been published, but he did not have the 5-volume set of the 1st edition. I sent him a complete set, he liked it and wanted to publish it again. That is how the 2nd edition was born. At first, I was very hesitant to take on such a large project. I am 81 years old with typical health problems of a senior citizen. I had always wished that the 1st edition could have been in chronological order, but that just was not possible when I was working on it during my "prime" years. But now I know the entire developmental span and can reorganize the 2nd edition in chronological order. Another problem with the 1st edition is that the books are large and heavy, difficult to carry around and consult in a busy laboratory. So, the 2nd edition is a series of 15 smaller books that can more easily travel if users want a print-on-demand paper copy. Finally, the 2nd edition will be entirely online so that users can access the books using laptops and tablets. Users will also have access to free extra materials for each specimen. These include videos of aligned low-magnification photos of all those taken for each specimen shown at 2 frames/second for a "fly through" of the head and neck in various cutting planes. Every section in the video is also available as a high-resolution jpeg image that can be downloaded. These new images have been scanned on a Nikon Coolscan Ved with a resolution of approximately 8,000 dpi. Every specimen, not just the ones we chose for the Atlas, is available on brainimages.org. Instead of playing Mahjong and going on cruises (dangerous for people my age anyway), I am glued to my computer checking labels, making slight revisions here and there, and reorganizing all the original plates. I hope I stay healthy to see it through!

Shirley A. Bayer
December 17, 2021

ACKNOWLEDGMENTS

We thank the late Dr. William DeMyer, pediatric neurologist at Indiana University Medical Center, for access to his personal library on human CNS development. We also thank the staff of the National Museum of Health and Medicine, who were at the Armed Forces Institute of Pathology, Walter Reed Hospital, Washington, D.C. when we collected data in 1995 and 1996: Dr. Adrianne Noe, Director; Archibald J. Fobbs, Curator of the Yakovlev Collection; Elizabeth C. Lockett; and William Discher. We are most grateful to the late Dr. James M. Petras at the Walter Reed Institute of Research, who made his darkroom facilities available so that we could develop all the photomicrographs on location rather than in our laboratory in Indiana. Finally, we thank Chuck Crumly, Neha Bhatt, Kara Roberts, Michele Dimont, and Rebecca Condit for expert help during production of the manuscript.

AUTHORS

Shirley A. Bayer received her PhD from Purdue University in 1974 and spent most of her scientific career working with Joseph Altman. She was a professor of biology at Indiana-Purdue University in Indianapolis for several years, where she taught courses in human anatomy and developmental neurobiology while continuing to do research in brain development. Her lengthy publication record of dozens of peer-reviewed scientific journal articles extends back to the mid 1970s. She has co-authored several books and many articles with her late spouse, Joseph Altman. It was her research (published in *Science* in 1982) that proved that new neurons are added to granule cells in the dentate gyrus during adult life, a unique neuronal population that grows. That paper stimulated interest in the dormant field of adult neurogenesis.

Joseph Altman, now deceased, was born in Hungary and migrated with his family via Germany and Australia to the United States. In New York, he became a graduate student in psychology in the laboratory of Hans-Lukas Teuber, earning a PhD in 1959 from New York University. He was a postdoctoral fellow at Columbia University, and later joined the faculty at the Massachusetts Institute of Technology. In 1968, he accepted a position as a professor of biology at Purdue University. During his career, he collaborated closely with Shirley A. Bayer. From the early 1960s to 2016, he published many articles in peer-reviewed journals, books, monographs, and online free books that emphasized developmental processes in brain anatomy and function. His most important discovery was adult neurogenesis, the creation of new neurons in the adult brain. This discovery was made in the early 1960s while he was based at MIT and was largely ignored in favor of the prevailing dogma that neurogenesis is limited to prenatal development. After Dr. Bayer's paper proved that new neurons are adding to granule cells in the hippocampus, his monumental discovery became more accepted. During the 1990s, new researchers "rediscovered" and confirmed his original finding. Adult neurogenesis has recently been proven to occur in the dentate gyrus, olfactory bulb, and striatum through the measurement of Carbon-14—the levels of which changed during nuclear bomb testing throughout the 20th century—in postmortem human brains. Today, many laboratories around the world are continuing to study the importance of adult neurogenesis in brain function. In 2011, Dr. Altman was awarded the Prince of Asturias Award, an annual prize given in Spain by the Prince of Asturias Foundation to individuals, entities, or organizations from around the world who make notable achievements in the sciences, humanities, and public affairs. In 2012, he received the International Prize for Biology, an annual award from the Japan Society for the Promotion of Science (JSPS) for "outstanding contribution to the advancement of research in fundamental biology." This prize is one of the most prestigious honors a scientist can receive. Dr. Altman died in 2016, and Dr. Bayer continues the work they started over 50 years ago. In his honor, she has set up the Altman Prize, awarded each year to an outstanding young researcher in developmental neuroscience by JSPS.

PART 1: INTRODUCTION

ORGANIZATION OF THE ATLAS

This is the first volume in the *Atlas of Human Central Nervous System Development* series, 2nd edition. It deals with human brain development during the early first trimester. The 4 specimens in this book have crown-rump (CR) lengths from 3.5- to 4.5-mm with estimated gestation weeks (GW) from 3.2 to 4.5. These specimens were the last four in Volume 5 of the 1st edition (Bayer and Altman, 2002-2008). The second edition of the *Atlas* is completely reorganized to present the brains of the most immature first trimester specimens in the first volume and ending with the third trimester (Volumes 2-13). The project finishes with the spinal cord (Volumes 14-15). The specimens presented here have been implanted in the uterine wall for only a few days to a week. The annotations emphasize four developmental processes: *first,* growth of the stockbuilding neuroepithelium along the expanding shorelines of the brain's protoventricles, *second,* early neurogenesis, *third,* the initial expansion of the superarachnoid reticulum, and *fourth,* the interactions between the brain and peripheral structures in the head and pharyngeal arches, especially as they relate to the rhombomeres and sensory cranial nerves.

The present volume features four normal specimens. Two are cut in the frontal/horizontal plane, two in the sagittal plane. Each specimen is presented as a series of grayscale photographs of its Nissl-stained nervous system sections including the surrounding body (**Parts II** through **V**). The photographs are shown from anterior to posterior (frontal/horizontal specimens) and medial to lateral (sagittal specimens). The dorsal part of each frontal/horizontal photo is toward the top of the page, the ventral part at the bottom, and the midline is in the vertical center. Both frontal/horizontal specimens have computer-aided 3-dimensional reconstructions of their brains showing each section's location. That reconstruction clears up the ambiguity about the exact plane of sectioning through each specimen. For all sagittal specimens, the left side of each photo is facing anterior, right side posterior, top side dorsal, and bottom side ventral.

SPECIMENS AND COLLECTIONS

The four specimens in this book are from the *Minot and Carnegie Collections* in the National Museum of Health and Medicine that used to be housed at the Armed Forces Institute of Pathology (AFIP) in Walter Reed Hospital in Washington, D.C. Since the AFIP closed, the National Museum was moved to Silver Springs, MD; these Collections are still available for research.

Three of the specimens are from the *Carnegie Collection* (designated by a **C** prefix), that started in the Department of Embryology of the Carnegie Institution of Washington. It was led by Franklin P. Mall (1862-1917), George L. Streeter (1873-1948), and George W. Corner (1889-1981). These specimens were collected during a 40– to 50– year time span and were histologically prepared with a variety of fixatives, embedding media, cutting planes, and histological stains. Early analyses of specimens were published in the early 1900s in *Contributions to Embryology, The Carnegie Institute of Washington* (now archived in the Smithsonian Libraries). O'Rahilly and Müller (1987, 1994) have given overviews of first trimester specimens in this collection.

One specimen is from the *Minot Collection* (designated by an **M** prefix), which is the work of Dr. Charles Sedgwick Minot (1852-1914), an embryologist at Harvard University. Throughout his career, Minot collected about 1900 embryos from a variety of species. The 100 human embryos were probably acquired between 1900 and 1910. From our examination of these specimens and their similar appearance, we assume that they are preserved in the same way, although we could not find any records describing fixation procedures. The slides contain information on section numbers, section thickness (6 μ, to 10 μ), and stain (aluminum cochineal). To our knowledge, we are the first to examine the nervous system in these specimens.

At this early stage, accurate crown-rump (CR) lengths and gestation week (GW) estimates are elusive. The freshly extracted specimens are barely visible and difficult to measure without damage. Menstrual age, used to estimate gestation weeks, is intrinsically variable because many women have irregular menstrual cycles. Since the publication of the *Atlas* 1st edition, we came across an article (Loughna et al., 2009) with ultrasound data that helped us to better estimate gestational age, so the ages in the 2nd edition will differ slightly. An example of the problem is in this volume. The most immature specimen, M714, has a 4.0-mm CR but is morphologically younger than C7724 with a 3.5-mm CR. The neural tube is still open anteriorly in M714 and is probably open at the tip of the spinal cord. M714 also has a slit-like ventricular system in the forebrain, midbrain, and spinal cord; only the rhomben-cephalon (hindbrain) has a slightly larger ventricle. Its CR would prompt us to assume that it is GW4, but it is more likely GW3.2. In contrast, C7724 has a completely closed neural tube and has larger ventricles than M714.

PLATE PREPARATION

Sections throughout the entire specimen were photographed in serial order with Kodak technical pan black-and-white negative film (#TP442). The film was developed for 6 to 7 minutes in a dilution F of Kodak HC-110 developer, a 30 second stop bath, Kodak fixer for 5 minutes, Kodak hypo-clearing agent for 1 minute, running water rinse for 10 minutes, and a brief rinse in Kodak photo-flo before drying. All sections of a given specimen were photographed at the same magnification. M714 was photographed with an Olympus microscope, C7724 and C9297 with a Wild Makroscop, and C836 with a stereozoom microscope.

The negatives were scanned at 2700 dots per inch (dpi) with a Nikon Coolscan-1000 35 mm negative film scanner attached to a Macintosh PowerMac G3 computer which had a plug-in driver built into Adobe Photoshop. The negatives were scanned as color positives because that brought out more subtle shades of gray. The original scans were converted to 300 dpi using the non-resampling method for image size. The powerful features of Adobe photoshop were used to enhance contrast, correct uneven staining, and slightly darken or lighten areas of uneven exposure.

The photos chosen for annotation in **Parts II** through **V** are presented as companion plates **A** and **B** on facing pages. **Part A** on the left shows the full-contrast photograph with labels of peripheral neural structures; **part B** on the right shows low-contrast copies of the same photograph with superimposed outlines of the labeled brain parts. The *low-magnification plates* show entire sections to identify the large structures and subdivisions of the brain. The few *high-magnification plates* in the two sagittal specimens feature enlarged views of the brain to show tissue organization. This type of presentation allows a user to see the entire section as it would appear in a microscope and then consult the detailed markup in the low-contrast copy on the facing page leaving little doubt about what is being identified. The labels themselves are not abbreviated, so the user does not need to consult a list. Different fonts are used to label different classes of structures: the ventricular system is labeled in **CAPITALS**, the neuroepithelium and other germinal zones in **Helvetica bold**, transient structures in ***Times bold italic***, and permanent structures in Times Roman or **Times bold**. Adobe Illustrator was used to superimpose labels and to outline structural details on the low-contrast images. Plates were placed into a book layout using Adobe InDesign. Finally, high-resolution portable document files (pdf) were uploaded to CRC Press/ Taylor & Francis websites.

3-DIMENSIONAL COMPUTER RECONSTRUCTIONS

This process took five steps. *First,* image files in the series for each specimen were placed into a Photoshop stack with each image in a separate layer. *Second,* by altering the transparency of these layers, the sections were aligned to each other. After alignment, each layer was exported as a separate file. *Third,* Adobe Illustrator was used to outline the brain surface of each aligned section, and these contours were saved in separate Adobe Illustrator encapsulated postscript (eps) files. *Fourth,* the eps files were imported into 3D space (x, y, and z coordinates) using Cinema 4DXL (C4D, Maxon Computer, Inc.). For each section, points on the contours have unique x-y coordinates and the same z coordinate. By calculating the distance between sections, the entire array of contours was stretched out in the z axis. The C4D loft tool builds a spline mesh of polygons starting with the x-y points on the contour with the most anterior z coordinate and ending with the x-y points on the contour with the most posterior z coordinate. The spline meshes of the entire brain surface were rendered completely opaque at various camera angles using the C4D ray-tracing engine (**Figures 6 to 9** in **Parts II** through **V**). *Fifth,* in all frontal/horizontal-sectioned specimens, models of the brain surface posterior to a specific section were rendered with a copy of the photograph of that section texture mapped as a front cap on the model (*insets* in **Part A**).

IDENTIFICATION OF TRANSIENT AND IMMATURE BRAIN REGIONS

Except for the major brain vesicles and the rhombomeres in the pons and medulla, the identification of most structures in early first trimester human brain—like the various neuroepithelial (NEP) compartments—have received little attention in the recent past. Most identifications in this volume are based on our previous experimental ^{3}H-thymidine autoradiographic work with rats (Bayer et al., 1993, 1995, Bayer and Altman, 2012-present). There is a great morphological similarity between the rat brain and human brain at this early stage of development (Bayer et al., 1993, 1995).

His (1886) was one of the first scientists to examine embryonic specimens. He made major contributions to histology, like inventing the microtome and using serial sections to reconstruct 3-D models of developing embryos. Indeed, the same techniques are used in this *Atlas*. His was able to describe the outgrowth of fibers from large motor neurons in the developing spinal cord and he tracked their growth into the extremities (Louis and Stapf, 2001). His and other early embryologists distinguished three components of the embryonic spinal cord (**Fig. 1**): the proliferative ependymal layer surrounding the spinal canal, the cell-rich mantle layer, and the cell-sparse marginal layer. But well over 100 years of embryonic research has given us a better understanding of embryonic tissue. **Figure 2** shows the terminology we use in this *Atlas* applied to the midbrain tegmental wall in a methacrylate-embedded rat embryo 11

FIGURE 1: TERMINOLOGY OF HIS AND OUR MODIFICATIONS

A. Transverse section of the spinal cord in a 4-week human embryo (after His)

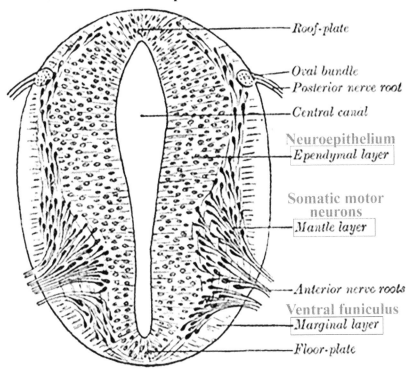

Roof-plate

Oval bundle
Posterior nerve root

Central canal

Neuroepithelium
Ependymal layer

Somatic motor neurons
Mantle layer

Anterior nerve roots

Ventral funiculus
Marginal layer

Floor-plate

The boxed labels are not used in this atlas. Rather, we replace those labels with the red ones next to each box. Our policy is to name components in the mantle and marginal layers as soon as they are identifiable. In this case, the mantle layer label is pointing to somatic motor neurons. The marginal layer label is pointing to the ventral funiculus.

B. Detail of the spinal cord ependymal layer in a 4-week human embryo (after His)

Germinal cell — Neuroepithelial cell nucleus in mitotic phase

Neuroblast — Neuroepithelial cell nucleus in synthetic phase

Nuclei of spongioblasts

Syncytium — All cells have individual, separate processes. There is no syncytium.

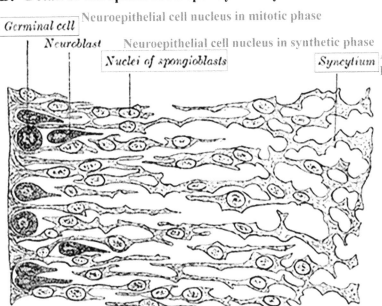

His did excellent detailed drawings of the developing nervous system and its various lamina. However, during his time, fixation, embedding media, and microtomes (which he invented) were unable to produce histological material that could decipher the very fine details of tissue organization. He thought that the neuroepithelium (ependymal layer) was a multi-layered structure. Later research showed it to be a pseudostratified columnar epithelium where nuclei migrate to and fro within the cytoplasmic tube.

https://theodora.com/anatomy/development_of_the_nervous_system.html

days after conception. The neuroepithelium (NEP) replaces the ependymal layer as shown in **Figure 1**. The nucleus migrates to the apical part of the cell to divide and migrates to the basal part of the cell to synthesize DNA for a new set of chromosomes. Note in **Figure 2** that mitotic figures are mainly at the ventricular edge (the apical part of the NEP). In 1935, Sauer provided evidence that this to and fro intranuclear migration existed, and the ³H-thymidine-auto-radiographic evidence (**Fig. 3**) proves that he is correct. An autoradiogram from an E15 rat killed 2 hours after its mother was injected with ³H-thymidine shows many heavily labeled nuclei in the basal part (synthetic zone) of the NEP.

The term *mantle layer* is not used for the cell-rich part of the embryonic nervous system outside the NEP. The

THE MIDBRAIN WALL IN A METHACRYLATE-EMBEDDED RAT EMBRYO ON E11

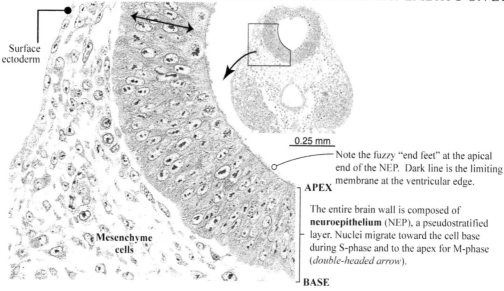

Surface ectoderm

Mesenchyme cells

0.25 mm

Note the fuzzy "end feet" at the apical end of the NEP. Dark line is the limiting membrane at the ventricular edge.

APEX

The entire brain wall is composed of **neuroepithelium** (NEP), a pseudostratified layer. Nuclei migrate toward the cell base during S-phase and to the apex for M-phase (*double-headed arrow*).

BASE

Figure 2. The fine structure of the brain wall lining in the future midbrain of a methacrylate-embedded rat on embryonic day (E)11. Methacrylate embedding preserves excellent histological detail in this 3μ section. Polymorphic mesenchyme cells that originated from the neural crest and mesoderm migrations are scattered between the brain and the surface ectoderm. (https://braindevelopmentmaps.org/home/methacrylate-brain-map-sets/e11-coronal-archive/)

FRONTAL SECTION OF AN E15 RAT TELENCEPHALON
2 HOURS AFTER EXPOSURE TO ³H-THYMIDINE

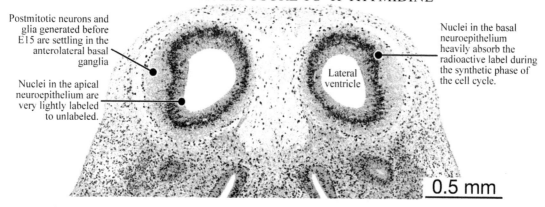

Postmitotic neurons and glia generated before E15 are settling in the anterolateral basal ganglia

Nuclei in the apical neuroepithelium are very lightly labeled to unlabeled.

Lateral ventricle

Nuclei in the basal neuroepithelium heavily absorb the radioactive label during the synthetic phase of the cell cycle.

0.5 mm

Figure 3. The autoradiographic proof of Sauer's observations in 1935. The injected ³H-Thymidine is absorbed by nuclei in the S (synthetic)-phase of the cell cycle in the basal part of the neuroepithelium. The very lightly labeled nuclei in the M (mitotic)-phase are at the ventricular edge and represent cells near the end of the S-phase that absorbed a small bit of labeled thymidine and migrated to the apex of the cell to divide between the time of injection and the time that this specimen was killed by immersion in Bouin's fixative. (6μ paraffin section, hematoxylin and eosin stain; https://braindevelopmentmaps.org/home/brain-map-sets/archived-2hr-survival-images/e15-2hr-survival-archived-images-coronal/)

cells migrating from the NEP are specific neurons moving in different directions, settling in various locations, and some are differentiating (*see* notes in **Fig. 1**). Extensive experimental evidence in the developing rat nervous system (reviewed in Bayer and Altman, 1995) enables identification of some of these neurons. Throughout the *Atlas of Prenatal Rat Brain Development* (Altman and Bayer, 1995), clumps of settling and migrating neurons in the parenchyma of the brain and spinal cord are named as soon as they are recognizable.

Evidence for two additional developmental structures is in **Figure 4**. *First*, the formation of the *superarachnoid reticulum* is related to parenchymal expansion of the brain (Bayer and Altman, 2002-2008). The superarachnoid reticulum is a broad, fluid-rich meningeal tissue sandwiched between the early-developing pia and dura. The superarachnoid reticulum appears before the brain parenchyma expands. As the brain parenchyma continually grows when neurons move in, settle, and differentiate, the superarachnoid reticulum continually shrinks until it is a thin meninx. We postulate that the transient hypertrophy of the superarachnoid reticulum serves as a *parenchymal expansion field* for the developing brain. The superarachnoid reticulum will be a prominent feature of all specimens in the first trimester. *Second*, the *boundary caps* mark spots on the pia where spinal nerves enter and exit the spinal cord and nerve entry or exit points in the brain (Bayer and Altman 2012-present). Boundary caps are important developmental structures and may be germinal zones for the Schwann cells that will surround peripheral nerves. The pia meninx is modified in these spots to allow axons to enter and exit the brain.

The term *neuroepithelium* deserves more discussion. It has been popular to call it the *ventricular zone* since 1970, mainly to describe the neuroepithelium that will generate cells of the cerebral cortex in the forebrain (Boulder Committee, 1970). We continue to use *neuroepithelium* throughout all volumes of this Atlas because it more accurately names the primary germinal matrix that produces cells specific to the nervous system: neurons, glia, choroid plexus, secondary germinal matrices, and finally, the ependymal lining of the ventricles.

In the four specimens presented here, the entire embryonic central nervous system is little more than a NEP continuum stretching from the forebrain to the tip of the spinal cord and actively growing in what we call a *stockbuilding* stage. Those who study cell lineage in the neuroepithelium call this the proliferative stage (Alvarez-Buylla, et al. 2001). During this stage, the NEP continuum expands as the fluid in the protoventricles increases; most NEP cells are not producing neurons, but rather are building up their numbers of progenitor cells that will later give rise to the many different neuronal populations and associated glia throughout the central nervous system.

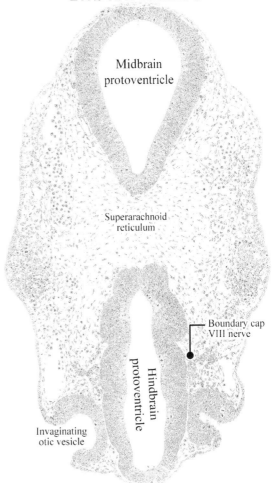

E11.5 RAT EMBRYO

Figure 4. A horizontal section through the midbrain and hindbrain of an E11.5 methacrylate-embedded rat embryo showing the cell-sparse **superarachnoid reticulum** in the cleft between the midbrain and hindbrain. The **boundary cap** of the vestibulo-auditory nerve is indicated just above the invaginating otic vesicle placode. (https://braindevelopmentmaps.org/home/-methacrylate-brain-map-sets/e11-5-coronal-archive/)

As the NEP expands, the "shoreline" of the fluid-filled ventricular system lengthens in relation to the size of the neuronal populations being generated for different brain structures at specific times. The uneven rates of proliferation results in a variegated ventricular shoreline. The brain vesicles themselves arise from the uneven proliferation (His, 1886), which also produces the optic vesicle evagination and the rhombomeres. In the four specimens in this volume, the rhombencephalon occupies the largest part of the NEP in the brain and is the site where some of the earliest-generated neurons will reside in the mature brain/spinal cord. As time goes on, the rhombencephalon will become a relatively smaller part of the brain. The telencephalon has barely emerged in the last two specimens and will continue to expand throughout the rest of development and into the early adult period. The human neocortical NEP has the longest stockbuilding stage (months) because of the enormous number of neurons that it will generate. It will occupy a larger and larger part of the brain during development and will form the largest structure in the mature brain—the neocortex. At the opposite extreme are some neuronal populations in the thalamus. Our autoradiographic work in rats shows that the stockbuilding stage takes several days, producing invaginations and evaginations of the third ventricular wall. Then, the NEP will massively unload neurons over a 24–48-hour period during the neurogenetic stage (Bayer and Altman, 2012-present) to generate specific thalamic nuclei.

The four specimens in this volume are equivalent to rat embryos on embryonic days (E) 11 to 11.5 based on our timetables of neurogenesis using ^{3}H-thymidine dating methods (Bayer and Altman, 2012-present). **Table 1** lists the few neuronal populations being generated. Many of these populations are still not distinguishable in the brain or spinal parenchyma, and often newly generated neurons are sequestered in the NEP before they migrate out (Bayer and Altman, 2012-present). The oldest cells in the central nervous system are the neurons in the mesencephalic nucleus of the trigeminal in the periphery of the midbrain central gray (over 80% generated on or before E11 in rats or around CRs 3.2-4.5 mm in humans). These neurons are one of two known populations that are generated in the peripheral nervous system (either or both from the neural crest and the trigeminal branchial placode) and the cell bodies migrate into the brain with the incoming fibers of the trigeminal nerve (V). Mesencephalic neurons migrating into the brain are shown in Plate 77, Volume 5, 1st edition, Bayer and Altman (2008). The other peripherally generated neurons migrate into the neurosecretory hypothalamus with the vomeronasal nerve and produce luteinizing hormone releasing hormone (Wray et al., 1989). The oldest centrally generated neurons are in the retrofacial nucleus in the respiratory center of the medulla (around 60% are generated on or before CRs from 3.5-4.5 mm).

Table 1. Populations Starting Neuro-genesis between CR 3.5-4.5 mm	
REGION	**NEURAL POPULATION**
SPINAL CORD	Cervical somatic motor neurons
MEDULLA	Hypoglossal (XII motor neurons)
	**Retrofacial nucleus
	Lateral vestibular nucleus
	Lateral trapezoid nucleus
	Raphe complex
PONS	Trigeminal (V) motor neurons
	Reticular formation
MIDBRAIN TEGMENTUM	Locus coeruleus
	**Mesencephalic nucleus (V)

**Indicates peak generation

The NEP continuum has the same microscopic appearance, but that similarity does not imply homogeneity. A most important concept is that the neuroepithelium from the very beginning is a *mosaic* where different parts will produce different neuronal populations. A piece of neuroepithelium from the anterior prosencephalon will develop into either basal ganglia or cerebral cortex. A piece from the roof of the midbrain will develop into either optic tectum (superior colliculus) or auditory tectum (inferior colliculus), while the midbrain floor (tegmentum) will develop into a variety of smaller structures, such as the motor neurons that control the external muscles of the eyeball and the red nucleus that will be part of feedback loops in cerebellar circuitry.

The hindbrain (rhombencephalon) has a series of 6 swellings, the rhombomeres, around the diamond-shaped ventricle (rhombus) that look like but are ***not*** segments in the proper sense of identical repeating units. We will give exhaustive evidence in this volume and several to follow that each rhombomere is associated with the sensory part of a cranial nerve that enters either the pons or medulla. Many placodes are tentatively identified as possible germinal sources for sensory cranial ganglia in the head and neck related to the rhombomeres (Fryer et al., 2011; Sudiwala and Knox, 2019; Som, 2020). The forebrain is dominated by the optic vesicle evagination. The olfactory placode will play a major role in specifying the neuroepithelium in the telencephalon. Thus, regional specification of the neuroepithelium appears to take many cues from placodes in the head, pharyngeal arches in the primitive mouth and throat, mesenchyme from neural crest migration, the invaginating otic vesicle (auditory and vestibular structures), and somites around the developing spinal cord. The somites are repeating blocks of tissue, but that is related to the necessity that the vertebral column has to be composed of small bones that allow twisting and turning movements of the skeletal system. The spinal cord itself is not segmented. Based on our observations, we cannot explain neural development using a segment-based terminology (prosomeres, neuromeres, etc., Puelles, 2001).

A most important developmental event is the neural crest migration (His, 1886; Bronner, 2012). That migration is another peripheral cue to induce neuroepithelial mosaicism. The neural crest migrates inward from the junction between the neural plate and the surface ectoderm mostly before the neural tube closes and ends just as the neural tube is closing. It is also possible that some neural crest cells delaminate from the evaginating optic vesicle. Mesenchyme derived from neural crest migration in the body form a variety of structures (in participation with mesenchyme from the mesoderm that migrates during gastrulation; Ghimire et al., 2021). Small patches of mesenchyme influence heterogeneity in the neuroepithelium. In the most immature specimen (**Part II**, M714), we were not fortunate to photograph the few sections where neural crest migration is still visible although the anterior neuropore is still open. As a substitute, we show the robust neural crest migration in a rat embryo on E10.5 (**Fig. 5**) just hours before the neural tube closes.

RAT EMBRYO ON E10.5 SHOWING NEURAL CREST MIGRATION

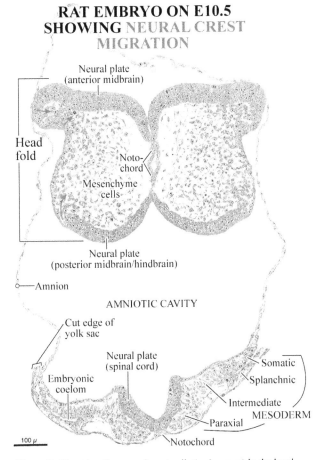

Figure 5. The migrating neural crest cells (*red arrows*) in the head fold of an E10.5 rat embryo invading the field of mesenchyme cells beneath the open neural plate (many are neural crest cells that migrated earlier). The neural crest migration is the result of a highly synchronized cascade of gene expression in the lateral edges of the neural plate to define the edge of the plate and to change the edge cells into a migratory population (Bronner, 2012). https://braindevelopment-maps.org/home/methacrylate-brain-map-sets/e10-5-transverse-archive/

Another feature shown in **Figure 5** is the notochord that stretches between the neural plate of the anterior and posterior midbrain head fold and is beneath the neural plate in the future spinal cord. The notochord appears during gastrulation as mesodermal cells migrate downward in the primitive streak and first fuse with the endoderm to form the notochordal plate (Som, 2020; Ghimire et al., 2021). Next, the plate separates from the endoderm at its fusion points and balls upward to form a tube, the notochord itself. Note that in **Figure 5**, the notochord label beneath the spinal neural plate is pointing to a plate-like structure rather than a rounded notochord. The notochord induces the formation of the neural plate in the midline ectoderm and sets up the ventral-to-dorsal axis in the embryonic nervous system. It always lies in the midline directly below the floor plate in the brainstem and spinal cord. The notochord secretes sonic hedgehog protein which influences the development of somatic motor neurons in the ventral neural tube throughout the spinal cord and up to the anterior extent of the midbrain. The prosencephalon does not have a notochord beneath it during development.

Finally, research has shown that specification of the NEP mosaic is basically the result of regional expression and suppression of specific genes and the accumulation of different proteins in NEP cells throughout various parts of the brain and spinal cord at specific times. Some of the fascinating ongoing research is reviewed by Bianchi (2018). I consulted the research she reviewed on *hox* gene expression in the various rhombomeres—(each one expresses a different set during development). Although none of the work presented in this *Atlas* is based on gene expression, the strictly timed neurogenetic events in the neuroepithelium and the morphological developmental sequences we show throughout this volume and the many volumes to come are directly dependent on the interplay of synchronized gene expression. The genetics underlying neural development has a rapidly expanding database that will no doubt eventually explain most of the spatiotemporal events we show in this entire series.

REFERENCES

A separate **Glossary** or a **Developmental Neuroscience Dictionary** is available to accompany this and all volumes in the second edition. Entries give brief definitions and developmental information on emerging structures in the nervous system. They are based on our previous research on nervous system development. Nearly all entries were also checked with searches on Google, PubMed, PubMed StatPearls, and Wikipedia.

Altman J, Bayer SA (1995) *An Atlas of Prenatal Rat Brain Development*. CRC Press, Boca Raton, Florida.

Alvarez-Buylla A, Garcia-Verdugo JM, Tramontin AD (2001) A unified hypothesis on the lineage of neural stem cells. *Nature Reviews Neuroscience*, 2, 287-292.

Bayer SA, Altman J, Russo RJ, Zhang X (1993) Timetables of neurogenesis in the human brain based on experimentally determined patterns in the rat. *Neurotoxicology* **14**: 83-144.

Bayer SA, Altman J, Russo RJ, Zhang X (1995) Embryology. In: *Pediatric Neuropathology*, Serge Duckett, Ed. Williams and Wilkins, pp. 54-107.

Bayer SA, Altman J (1995) Development: Some principles of neurogenesis, neuronal migration and neural circuit formation. In: *The Rat Nervous System*, 2nd Edition, George Paxinos, Ed. Academic Press, Orlando, Florida, pp. 1079-1098.

Bayer SA, Altman J (2002-2008) *Atlas of Human Central Nervous System Development* (First Edition), Volumes 1-5, CRC Press.

Bayer SA, Altman J (2008) *Atlas of Human Central Nervous System Development* (First Edition) Volumes 5, Appendix, CRC Press.

Bayer SA, Altman J (2012-present) www.neurondevelopment.org (This website has downloadable pdf files of our scientific papers on rat and human brain development grouped by subject.)

Bayer, SA (2013-present) www.braindevelopmentmaps.org (This website is a database containing high-resolution images of methacrylate-embedded normal rat embryos and of paraffin-embedded rat embryos exposed to ^{3}H-Thymidine.)

Bianchi LM (2018) *Developmental Neurobiology*, Garland Science, Taylor and Francis Group.

Boulder Committee (1970) Embryonic vertebrate central nervous system: Revised terminology. *Anat. Rec* 166, 257–261.

Bronner, ME, (2012) Formation and migration of neural crest cells in the vertebrate embryo. *Histochem. Cell Biol.*, 138, 179-186.

Freyer L, Aggarwal V, Morrow BE (2011) Dual embryonic origin of the mammalian otic vesicle forming the inner ear. *Development*, 138:5403-5414.

Ghimire S, Mantziou V, Moris N, Arias, AM (2021) Human gastrulation: The embryo and its models. *Developmental Biolgy*, 474:100-108.

Hamilton WJ, Boyd JD, Mossman HW (1964) *Human Embryology*, Revised 3rd Edition, Williams and Wilkins, Baltimore.

His W (1886) Zur Geschichte des menschlichen Rückenmarkes und der Nervenwurzeln: Sächsische Akademie der Wissenschaften zu Leipzig. *Mathematisch-Physicalische Classe.*1886;13:477-513.

Hochstetter F (1919) *Beiträge zur Entwicklungsgeschichte des menschlichen Gehirns*. Vol. 1. Leipzig und Wien: Deuticke.

Loughna P, Citty L, Evans T, Chudleigh T (2009) Fetal size and dating: Charts recommended for clinical obstetric practice, *Ultrasound*, 17:161-167.

Louis, ED Stapf, C (2001) Unraveling the neuron jungle, the 1879-1886 publications by Wilhelm His on the embryological development of the human brain, *Arch Neurol* 58, 1933-1935.

Meyer G, Gonzalez-Gómez (2018) The subpial granular layer and transient versus persisting Cajal-Retzius neurons of the human fetal cortex. *Cerebral Cortex*, 28:2043-2058.

O'Rahilly R; Müller F. (1987) *Developmental Stages in Human Embryos, Carnegie Institution of Washington*, Publication 637.

O'Rahilly R; Müller F. (1994) *The Embryonic Human Brain*, Wiley-Liss, New York.

Patton BM (1953) *Human Embryology*, 2nd Edition, McGraw-Hill, New York.

Puelles, I. (2001) Brain segmentation and forebrain development in amniotes. *Brain Research Bulletin*, 55, 695-710.

Sauer, FC (1935) Mitosis in the neural tube. *J Comp Neurol*, 62, 377-405.

Som PM (2020) The role of the placodes in the development of the glossopharyngeal, vagal, and trigeminal ganglia. *Neurographics*, 10:163-181.

Sudiwala S, Knox SM (2019) The emerging role of cranial nerves in shaping craniofacial development. *Genesis*, 57:e23282.

Wray S, Grant P, Gainer, H (1989) Evidence that cells expressing luteinizing hormone-releasing hormone mRNA in the mouse are derived from progenitor cells in the olfactory placode. *Proc. Natl Acad. Sci. USA*, 86:8132-8136.

PART II: M714
CR 4.0 mm (GW 3.2)
Frontal/Horizontal

This specimen is embryo #714 in the Minot Collection, designated here as M714. The crown-rump length (CR) is 4 mm. CR length is an unreliable measure to estimate gestational age because this specimen is much less mature than C7724 in Part 2, which also has a 4-mm CR. Since the number of somites could not be counted accurately in transverse sections, age determination is based on the degree of maturation of the central nervous system. The anterior neuropore closes at the 20-somite stage (Patten, 1953; Hamilton et al., 1959); M714 has a large open anterior neuropore. Using the timetables in Patten (1953) and Hamilton et al. (1959), we estimate that M714 has approximately 17 to 18 somites and is at gestational week (GW) 3.2. M714's prosencephalic and anterior mesencephalic sections are cut (8 μm) in the coronal plane, but the plane shifts to predominantly horizontal in the posterior mesencephalon, pons, and medulla. We photographed 21 sections at low magnification from the first section containing the head to the posterior tips of the rhombencephalon. Fifteen of these sections are illustrated in **Plates 1AB to 27AB**. All photographs were used to produce computer-aided 3-D reconstructions of the external features of M714's brain and optic vesicle (**Figure 6**), and to show each illustrated section *in situ* (*insets*, **Plates 1A to 13A**). Each illustrated section shows the brain with all surrounding tissues. Labels in **A Plates** (normal-contrast images) identify non-neural and peripheral neural structures; labels in **B Plates** (low-contrast images) identify central neural structures.

The prosencephalon is small and incomplete with a slit-like protoventricle that is open at the most anterior edge, the anterior neuropore. The neuropore is closing at the dorsal edge of these sections but is wide open ventrally (**Plate 1AB**). The sections in Plate 1 have a continuum between the neuroepithelium and the cephalic preplacodal ectodermal epithelium. At that juncture, the neural crest cells migrate into the space around the developing brain and become mesenchyme components that give rise to a variety of structures in the head and neck. The neural tube is closed in all the remaining sections that are analyzed. But it is likely that the most posterior sections contain an open posterior neuropore at the end of the spinal cord. There are large optic vesicle evaginations in back of the anterior neuropore that actually appear before the neural tube closes, but we did not analyze a human specimen younger than M714. This is shown in rats at embryonic days 10 and 10.5 (*see* braindevelopmentmaps.org). The fact that the neural tube is still open in M714 makes its ventricular system much smaller than all the other specimens in the first trimester. The largest part of the ventricular system is in the rhombencephalon (**Plates 11AB and 12AB**) which is farthest away from the open ends of the neural tube.

The mesencephalon contains a stockbuilding neuroepithelium surrounding a narrow keyhole-shaped protoventricle. The tectal neuroepithelium is a small arch over the top, while the tegmental and isthmal neuroepithelia form the slite-shaped bottom. There is a very thin cell-sparse primordial plexiform layer in future tegmental and isthmal areas.

The most prominent neuroepithelial structures in the rhombencephalon are the rhombomeres. The swellings represent areas of increased cell division in the neuroepithelium. In this specimen and all the others in this volume, the rhombomeres are associated with sensory cranial ganglia in the pons and medulla. The trigeminal ganglion is associated with rhombomere 2, the facial ganglion with rhombomere 3, the vestibulocochlear ganglion with rhombomeres 4 and 5, the glossopharyngeal ganglia with rhombomere 6, the vagal ganglia with rhombomere 7.

The cerebellar neuroepithelium is directly adjacent to rhombomere 2 in the most posterior sections where the dorsal rhombencephalic neuroepithelium blends with tectal/isthmal neuroepithelia, prompting earlier anatomists to call the cerebellum the first rhombomere.

M714 Computer-aided 3-D Brain Reconstructions

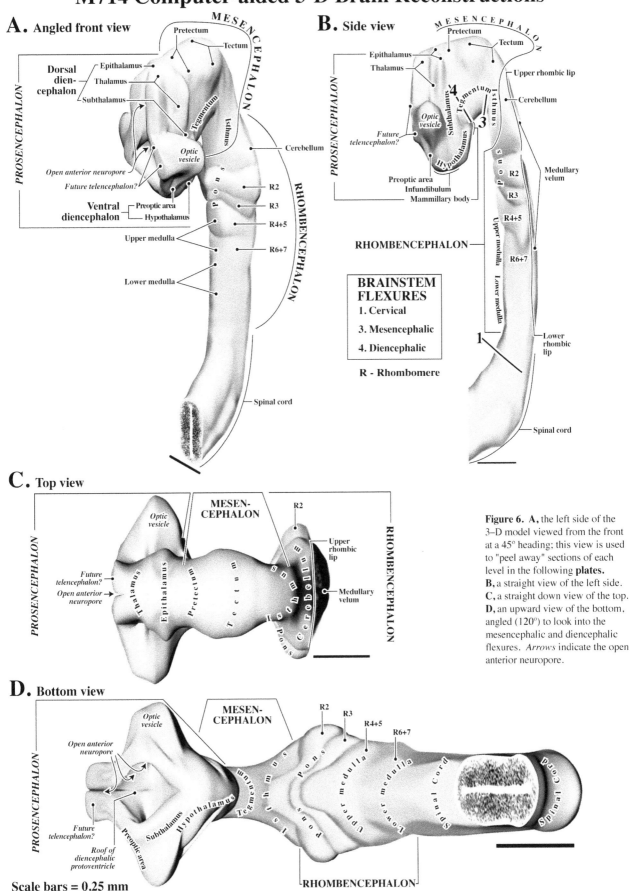

Figure 6. A, the left side of the 3–D model viewed from the front at a 45° heading; this view is used to "peel away" sections of each level in the following **plates.** **B,** a straight view of the left side. **C,** a straight down view of the top. **D,** an upward view of the bottom, angled (120°) to look into the mesencephalic and diencephalic flexures. *Arrows* indicate the open anterior neuropore.

A. Angled front view

Pretectum

MESENCEPHALON

Tectum

Dorsal diencephalon
— Epithalamus
— Thalamus
— Subthalamus

Tegmentum

Isthmus

PROSENCEPHALON

Optic vesicle

Cerebellum

Open anterior neuropore

Future telencephalon?

Ventral diencephalon
— Preoptic area
— Hypothalamus

Pons

R2

R3

R4+5

RHOMBENCEPHALON

Upper medulla

R6+7

Lower medulla

Spinal cord

B. Side view

MESENCEPHALON

Pretectum

Tectum

Epithalamus

Thalamus

Upper rhombic lip

4

Subthalamus

Tegmentum

Isthmus

Cerebellum

Optic vesicle

3

Future telencephalon?

Pons

Preoptic area

Hypothalamus

R2

R3

Infundibulum

Mammillary body

R4+5

Medullary velum

Upper medulla

Lower medulla

R6+7

RHOMBENCEPHALON

BRAINSTEM FLEXURES
1. Cervical
3. Mesencephalic
4. Diencephalic

R - Rhombomere

1

Lower rhombic lip

Spinal cord

C. Top view

MESENCEPHALON

Optic vesicle

R2

PROSENCEPHALON

Future telencephalon?

Open anterior neuropore

Thalamus

Epithalamus

Pretectum

Tectum

Upper rhombic lip

Cerebellum

Isthmus

Pons

Medullary velum

RHOMBENCEPHALON

D. Bottom view

Optic vesicle

MESENCEPHALON

Open anterior neuropore

R2

R3

R4+5

R6+7

PROSENCEPHALON

Future telencephalon?

Roof of diencephalic protoventricle

Preoptic area

Subthalamus

Hypothalamus

Tegmentum

Isthmus

Pons

Upper medulla

Lower medulla

Spinal Cord

Spinal Cord

RHOMBENCEPHALON

Scale bars = 0.25 mm

12

PLATE 1A

CR 4.0 mm
GW 3.2
M714
Frontal/
Horizontal

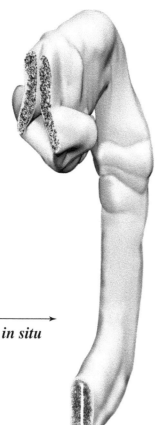

Section 3 brain *in situ*

Section 3

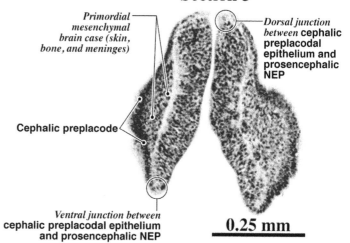

Primordial mesenchymal brain case (skin, bone, and meninges)

Dorsal junction between cephalic preplacodal epithelium and prosencephalic NEP

Cephalic preplacode

Ventral junction between cephalic preplacodal epithelium and prosencephalic NEP

0.25 mm

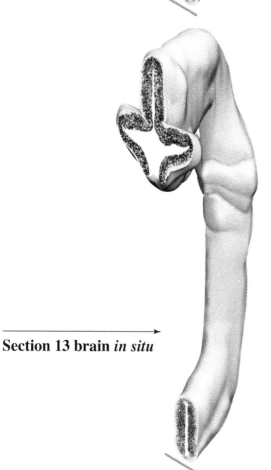

Section 13 brain *in situ*

Section 13

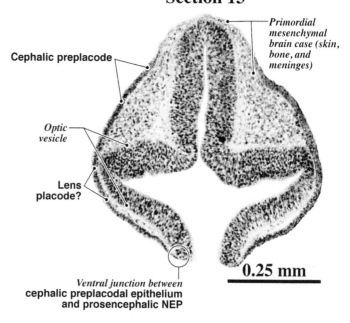

Cephalic preplacode

Primordial mesenchymal brain case (skin, bone, and meninges)

Optic vesicle

Lens placode?

Ventral junction between cephalic preplacodal epithelium and prosencephalic NEP

0.25 mm

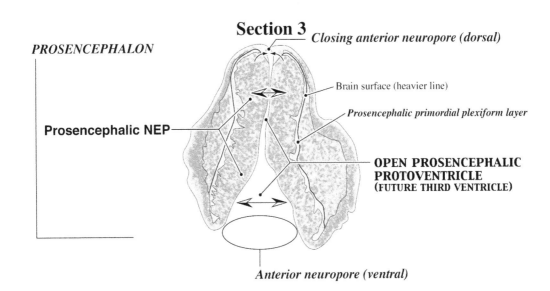

Section 3

PROSENCEPHALON

Prosencephalic NEP

Closing anterior neuropore (dorsal)

Brain surface (heavier line)

Prosencephalic primordial plexiform layer

OPEN PROSENCEPHALIC PROTOVENTRICLE
(FUTURE THIRD VENTRICLE)

Anterior neuropore (ventral)

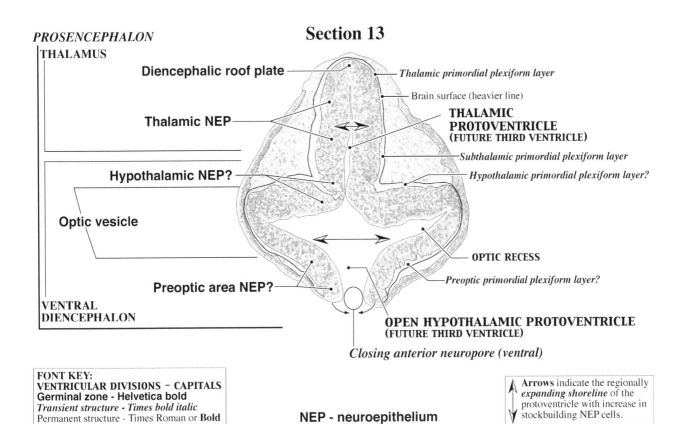

Section 13

PROSENCEPHALON
THALAMUS

Diencephalic roof plate

Thalamic NEP

Hypothalamic NEP?

Optic vesicle

Preoptic area NEP?

VENTRAL
DIENCEPHALON

Thalamic primordial plexiform layer

Brain surface (heavier line)

THALAMIC PROTOVENTRICLE
(FUTURE THIRD VENTRICLE)

Subthalamic primordial plexiform layer

Hypothalamic primordial plexiform layer?

OPTIC RECESS

Preoptic primordial plexiform layer?

OPEN HYPOTHALAMIC PROTOVENTRICLE
(FUTURE THIRD VENTRICLE)

Closing anterior neuropore (ventral)

FONT KEY:
VENTRICULAR DIVISIONS – CAPITALS
Germinal zone - Helvetica bold
Transient structure - Times bold italic
Permanent structure - Times Roman or **Bold**

NEP - neuroepithelium

Arrows indicate the regionally *expanding shoreline* of the protoventricle with increase in stockbuilding NEP cells.

14

**CR 4.0 mm, GW3.2
M714, Frontal/Horizontal
Section 18**

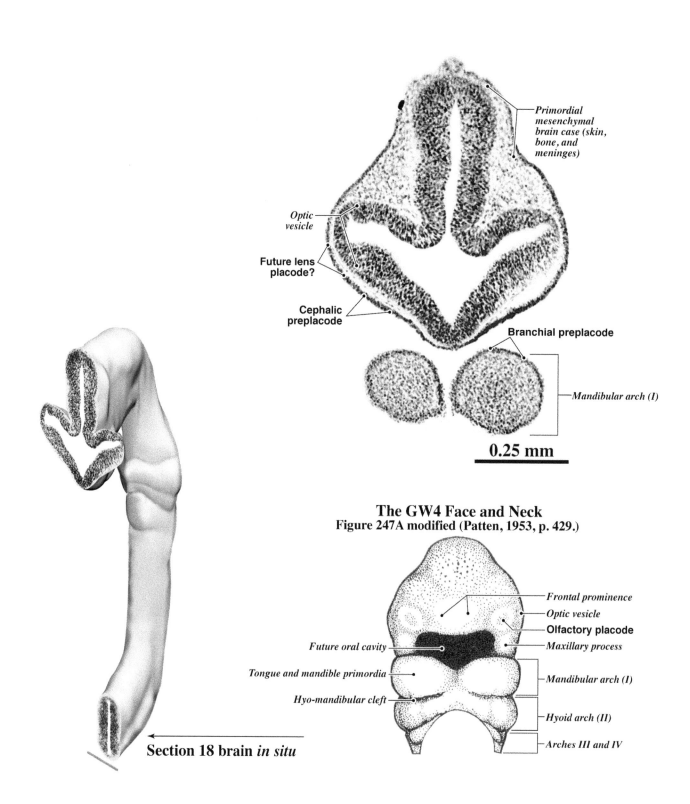

*Primordial
mesenchymal
brain case (skin,
bone, and
meninges)*

Optic
vesicle

Future lens
placode?

Cephalic
preplacode

Branchial preplacode

Mandibular arch (I)

0.25 mm

The GW4 Face and Neck
Figure 247A modified (Patten, 1953, p. 429.)

Frontal prominence
Optic vesicle
Olfactory placode
Maxillary process

Future oral cavity

Tongue and mandible primordia

Hyo-mandibular cleft

Mandibular arch (I)

Hyoid arch (II)

Arches III and IV

Section 18 brain *in situ*

DIENCEPHALON
  THALAMUS/EPITHALAMUS

Diencephalic roof plate
(pineal gland primordium?)

Epihalamic NEP?

Thalamic NEP?

SUBTHALAMUS
Subthalamic NEP

Hypothalamic NEP?

Future pigment
epithelium and retina

**Anterior
hypothalamic
NEP?**

Preoptic area NEP?

Diencephalic floor plate

**VENTRAL
DIENCEPHALON**

Thalamic/epithalamic primordial plexiform layer

Brain surface (heavier line)

Subthalamic primordial plexiform layer

Hypothalamic primordial plexiform layer

OPTIC RECESS

Hypothalamic primordial plexiform layer

Preoptic primordial plexiform layer

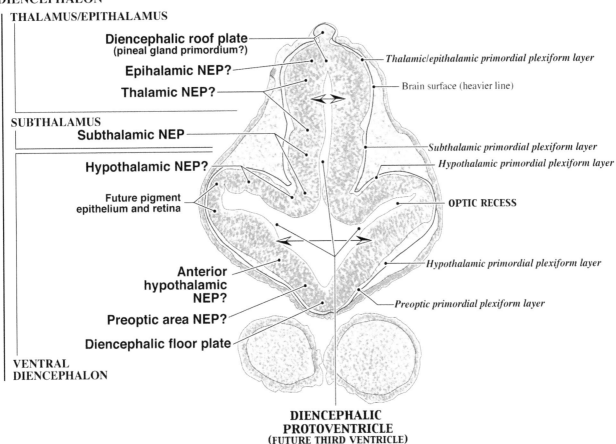

**DIENCEPHALIC
PROTOVENTRICLE**
(FUTURE THIRD VENTRICLE)

FONT KEY:
VENTRICULAR DIVISIONS – CAPITALS
Germinal zone - Helvetica bold
Transient structure - Times bold italic
Permanent structure - Times Roman or **Bold**

NEP - neuroepithelium

Arrows indicate the regionally
expanding shoreline of the
protoventricle with increase in
stockbuilding NEP cells.

PLATE 3A

CR 4.0 mm, GW3.2
M714, Frontal/Horizontal
Section 28

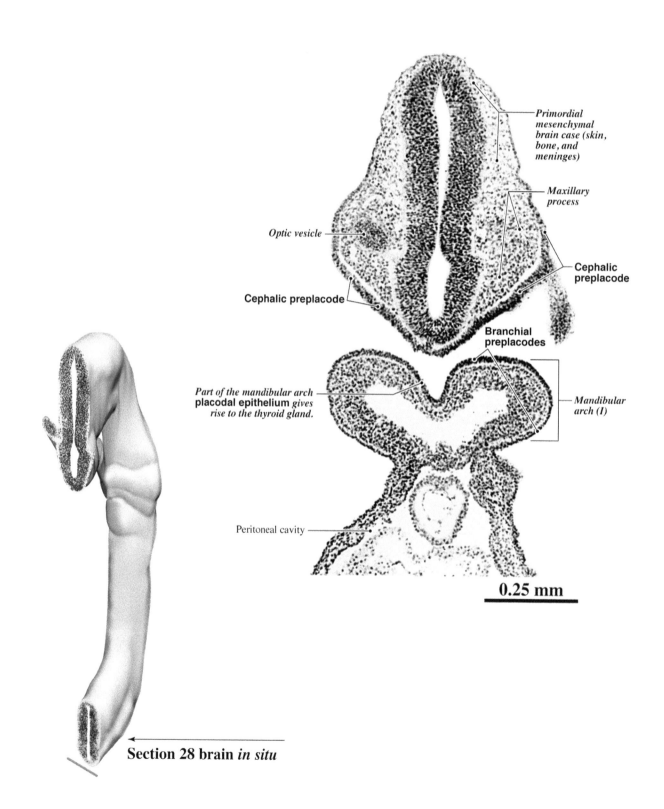

Primordial mesenchymal brain case (skin, bone, and meninges)

Maxillary process

Optic vesicle

Cephalic preplacode

Cephalic preplacode

Branchial preplacodes

Part of the mandibular arch placodal epithelium *gives rise to the thyroid gland.*

Mandibular arch (I)

Peritoneal cavity

0.25 mm

Section 28 brain *in situ*

Central neural structures labeled

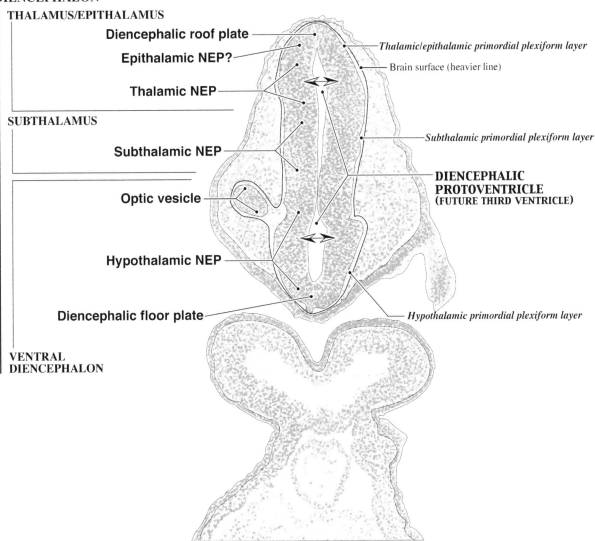

DIENCEPHALON

THALAMUS/EPITHALAMUS

Diencephalic roof plate

Epithalamic NEP?

Thalamic NEP

SUBTHALAMUS

Subthalamic NEP

Optic vesicle

Hypothalamic NEP

Diencephalic floor plate

VENTRAL DIENCEPHALON

Thalamic/epithalamic primordial plexiform layer

Brain surface (heavier line)

Subthalamic primordial plexiform layer

DIENCEPHALIC PROTOVENTRICLE (FUTURE THIRD VENTRICLE)

Hypothalamic primordial plexiform layer

FONT KEY:
VENTRICULAR DIVISIONS – CAPITALS
Germinal zone - Helvetica bold
Transient structure - Times bold italic
Permanent structure - Times Roman or **Bold**

NEP - neuroepithelium

Arrows indicate the regionally *expanding shoreline* of the protoventricle with increase in stockbuilding NEP cells.

18

**CR 4.0 mm, GW3.2
M714, Frontal/Horizontal
Section 33**

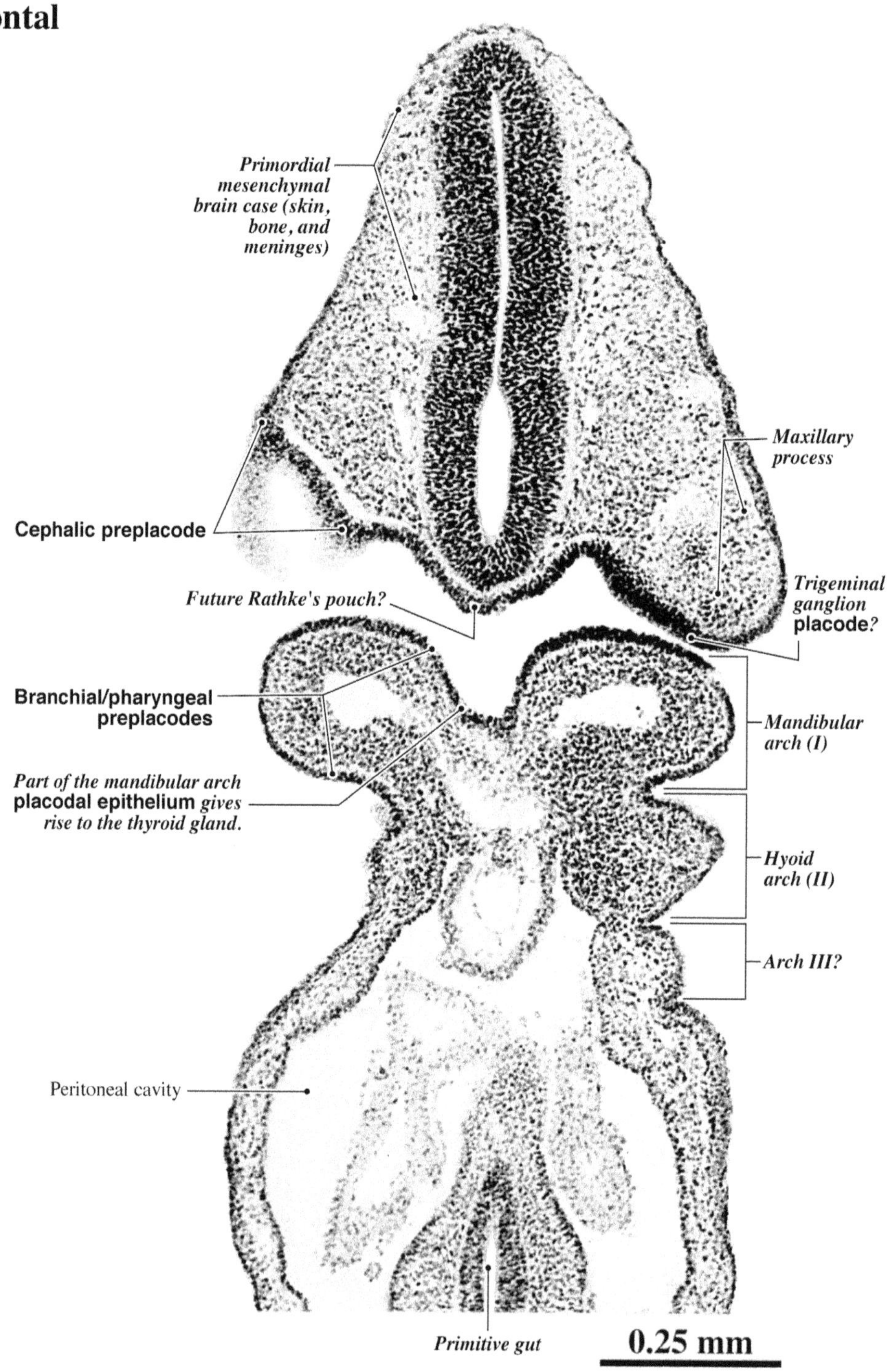

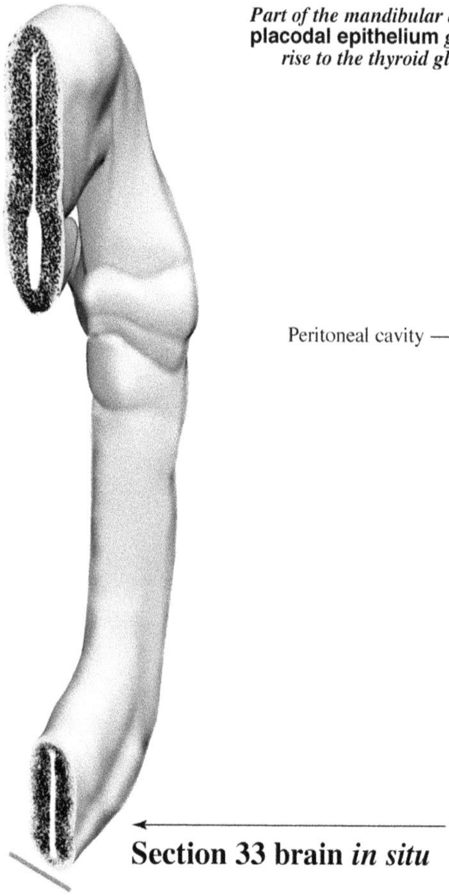

**Peripheral neural and
non-neural structures labeled**

*Primordial
mesenchymal
brain case (skin,
bone, and
meninges)*

*Maxillary
process*

Cephalic preplacode

*Trigeminal
ganglion
placode?*

Future Rathke's pouch?

Branchial/pharyngeal
preplacodes

*Mandibular
arch (I)*

Part of the mandibular arch
placodal epithelium *gives
rise to the thyroid gland.*

*Hyoid
arch (II)*

Arch III?

Peritoneal cavity

Primitive gut

0.25 mm

Section 33 brain *in situ*

Central neural structures labeled   **PLATE 4B**

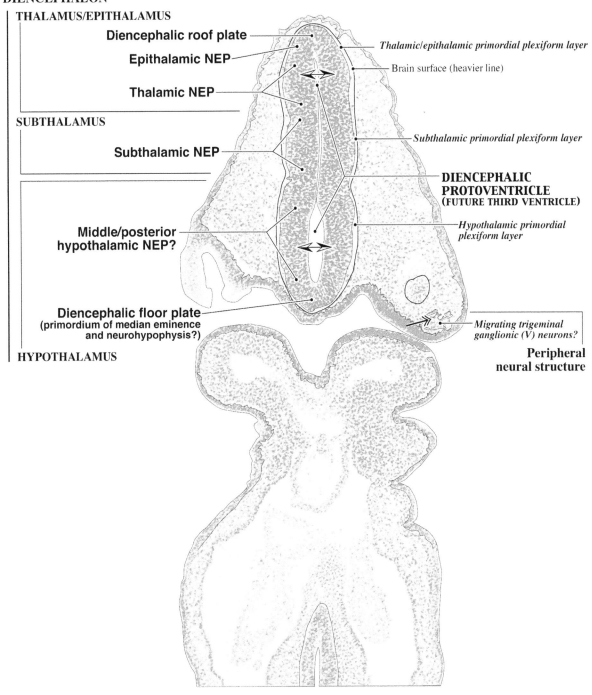

DIENCEPHALON

THALAMUS/EPITHALAMUS

Diencephalic roof plate ——————

Epithalamic NEP ——————

Thalamic NEP ——————

SUBTHALAMUS

Subthalamic NEP ——————

Middle/posterior
hypothalamic NEP? ——————

Diencephalic floor plate
(primordium of median eminence
and neurohypophysis?)

HYPOTHALAMUS

Thalamic/epithalamic primordial plexiform layer

Brain surface (heavier line)

Subthalamic primordial plexiform layer

**DIENCEPHALIC
PROTOVENTRICLE**
(FUTURE THIRD VENTRICLE)

*Hypothalamic primordial
plexiform layer*

*Migrating trigeminal
ganglionic (V) neurons?*

**Peripheral
neural structure**

FONT KEY:
VENTRICULAR DIVISIONS – CAPITALS
Germinal zone - Helvetica bold
Transient structure - Times bold italic
Permanent structure - Times Roman or **Bold**

NEP - neuroepithelium

Arrows indicate the
presumed *direction of
neuron migration* from
germinal sources.

Arrows indicate the regionally
expanding shoreline of the
protoventricle with increase in
stockbuilding NEP cells.

20

PLATE 5A

CR 4.0 mm, GW3.2
M714, Frontal/Horizontal
Section 38

Peripheral neural and
non-neural structures labeled

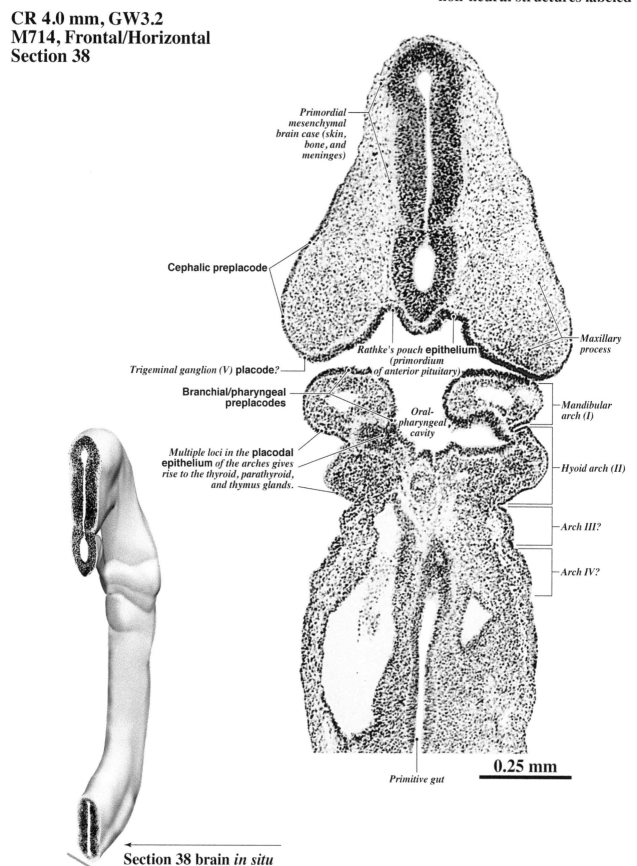

Primordial mesenchymal brain case (skin, bone, and meninges)

Cephalic preplacode

Trigeminal ganglion (V) placode?

Branchial/pharyngeal preplacodes

Multiple loci in the placodal epithelium *of the arches gives rise to the thyroid, parathyroid, and thymus glands.*

Rathke's pouch epithelium *(primordium of anterior pituitary)*

Maxillary process

Oral-pharyngeal cavity

Mandibular arch (I)

Hyoid arch (II)

Arch III?

Arch IV?

0.25 mm

Primitive gut

Section 38 brain *in situ*

Central neural structures labeled   **PLATE 5B**

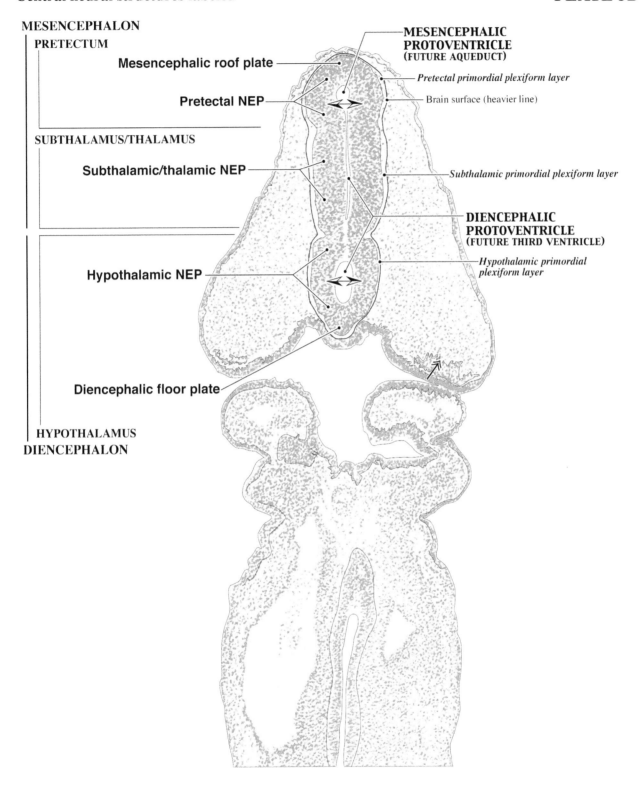

MESENCEPHALON
  PRETECTUM

MESENCEPHALIC roof plate

Pretectal NEP

SUBTHALAMUS/THALAMUS

Subthalamic/thalamic NEP

Hypothalamic NEP

Diencephalic floor plate

HYPOTHALAMUS
DIENCEPHALON

MESENCEPHALIC PROTOVENTRICLE (FUTURE AQUEDUCT)

Pretectal primordial plexiform layer

Brain surface (heavier line)

Subthalamic primordial plexiform layer

DIENCEPHALIC PROTOVENTRICLE (FUTURE THIRD VENTRICLE)

Hypothalamic primordial plexiform layer

FONT KEY:
VENTRICULAR DIVISIONS – CAPITALS
Germinal zone - Helvetica bold
Transient structure - Times bold italic
Permanent structure - Times Roman or **Bold**

NEP - neuroepithelium

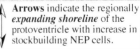

Arrows indicate the presumed *direction of neuron migration* from germinal sources.

Arrows indicate the regionally *expanding shoreline* of the protoventricle with increase in stockbuilding NEP cells.

PLATE 6A

CR 4.0 mm, GW3.2
M714, Frontal/Horizontal
Section 43

Peripheral neural
and non-neural
structures
labeled

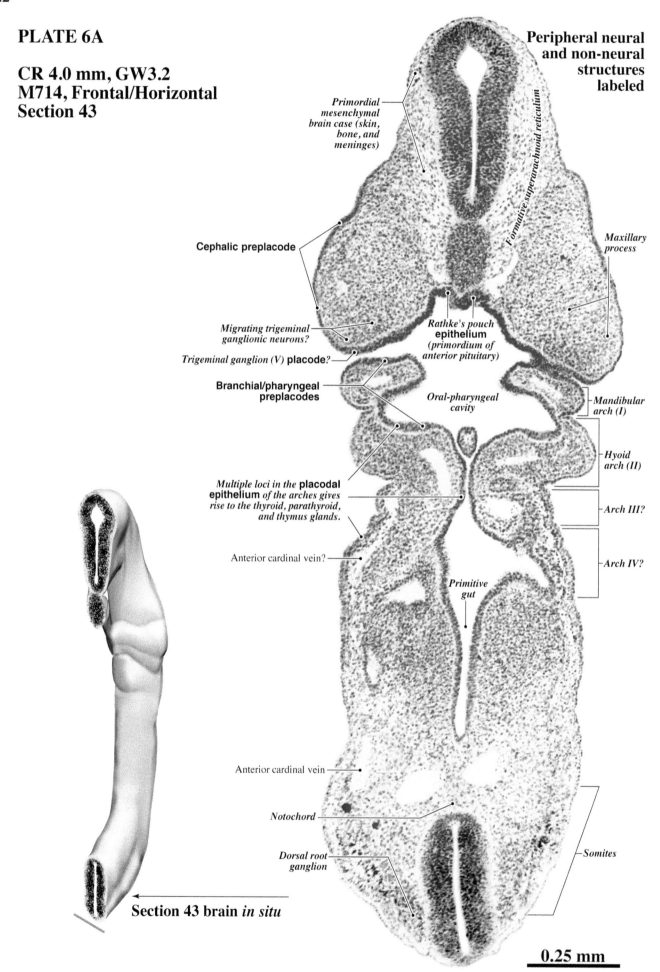

Primordial
mesenchymal
brain case (skin,
bone, and
meninges)

Formative superarachnoid reticulum

Maxillary
process

Cephalic preplacode

Rathke's pouch
epithelium
(primordium of
anterior pituitary)

Migrating trigeminal
ganglionic neurons?

Trigeminal ganglion (V) **placode?**

Branchial/pharyngeal
preplacodes

Oral-pharyngeal
cavity

Mandibular
arch (I)

Hyoid
arch (II)

Arch III?

Multiple loci in the **placodal**
epithelium *of the arches gives*
rise to the thyroid, parathyroid,
and thymus glands.

Arch IV?

Anterior cardinal vein?

Primitive
gut

Anterior cardinal vein

Notochord

Somites

Dorsal root
ganglion

Section 43 brain *in situ*

0.25 mm

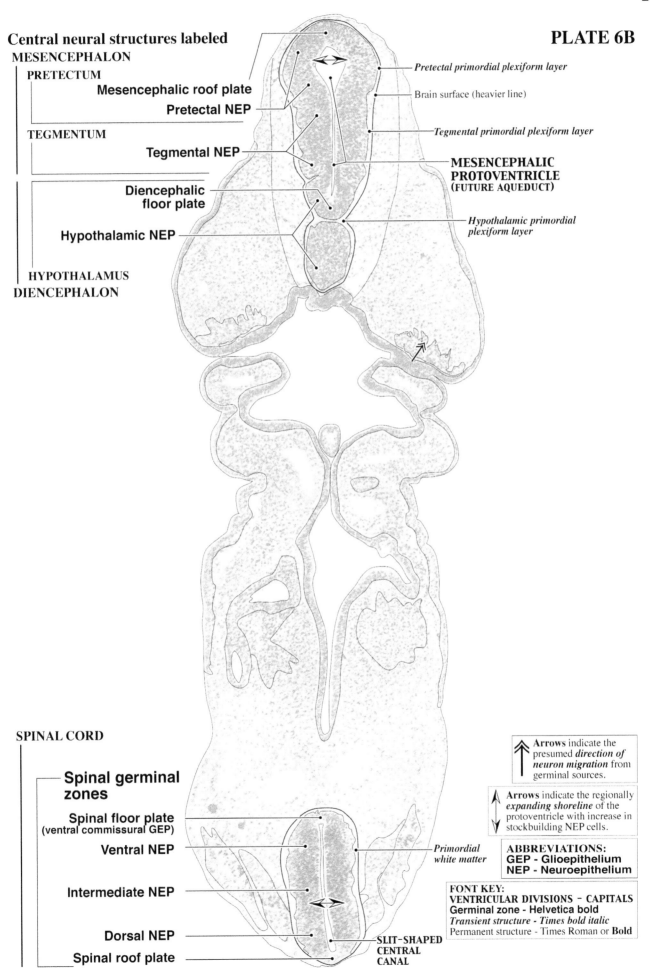

Central neural structures labeled
MESENCEPHALON
PRETECTUM
Mesencephalic roof plate
Pretectal NEP
TEGMENTUM
Tegmental NEP
Diencephalic floor plate
Hypothalamic NEP
HYPOTHALAMUS
DIENCEPHALON

Pretectal primordial plexiform layer

Brain surface (heavier line)

Tegmental primordial plexiform layer

MESENCEPHALIC PROTOVENTRICLE (FUTURE AQUEDUCT)

Hypothalamic primordial plexiform layer

SPINAL CORD

Spinal germinal zones

Spinal floor plate
(ventral commissural GEP)

Ventral NEP

Primordial white matter

Intermediate NEP

Dorsal NEP

Spinal roof plate

SLIT-SHAPED CENTRAL CANAL

Arrows indicate the presumed *direction of neuron migration* from germinal sources.

Arrows indicate the regionally *expanding shoreline* of the protoventricle with increase in stockbuilding NEP cells.

ABBREVIATIONS:
GEP - Glioepithelium
NEP - Neuroepithelium

FONT KEY:
VENTRICULAR DIVISIONS – CAPITALS
Germinal zone - Helvetica bold
Transient structure - Times bold italic
Permanent structure - Times Roman or **Bold**

24

PLATE 7A

CR 4.0 mm, GW3.2
M714, Frontal/Horizontal
Section 58

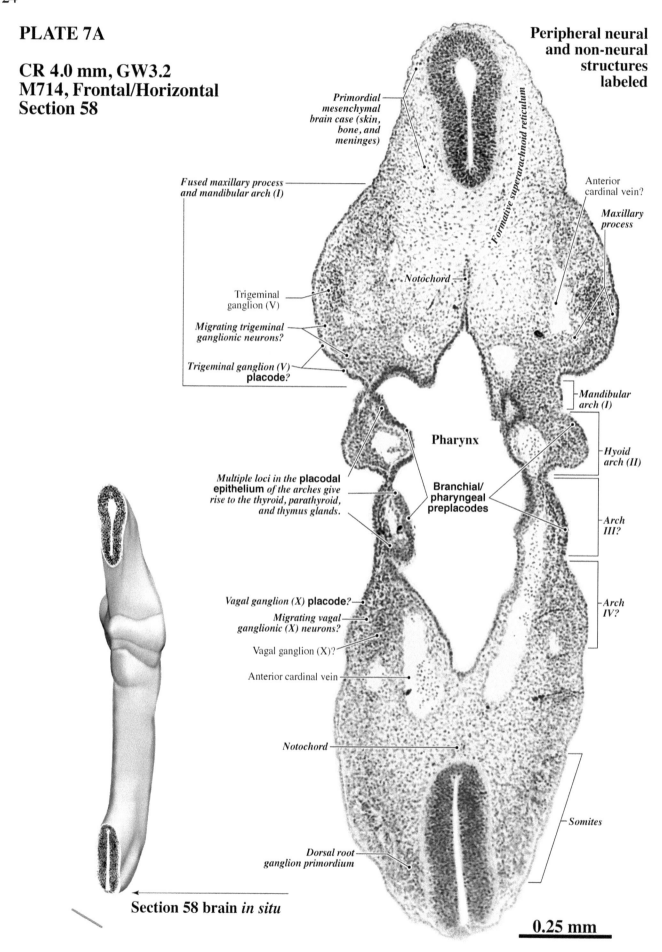

Peripheral neural
and non-neural
structures
labeled

Primordial mesenchymal brain case (skin, bone, and meninges)

Formative superarachnoid reticulum

Anterior cardinal vein?

Maxillary process

Fused maxillary process and mandibular arch (I)

Notochord

Trigeminal ganglion (V)

Migrating trigeminal ganglionic neurons?

Trigeminal ganglion (V) placode?

Mandibular arch (I)

Pharynx

Hyoid arch (II)

Multiple loci in the placodal epithelium *of the arches give rise to the thyroid, parathyroid, and thymus glands.*

Branchial/ pharyngeal preplacodes

Arch III?

Vagal ganglion (X) placode?

Migrating vagal ganglionic (X) neurons?

Vagal ganglion (X)?

Anterior cardinal vein

Arch IV?

Notochord

Somites

Section 58 brain *in situ*

Dorsal root ganglion primordium

0.25 mm

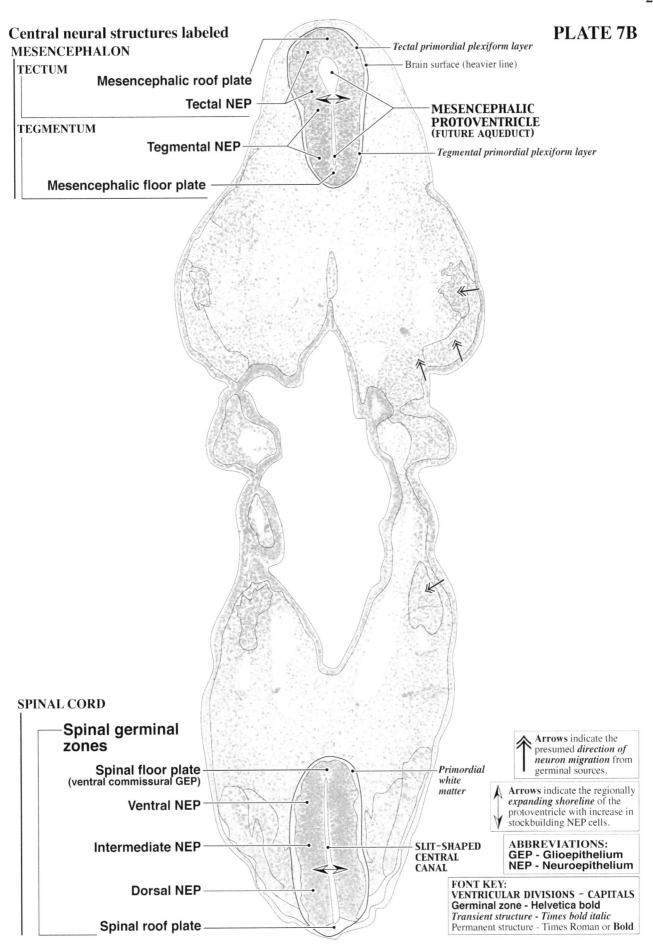

25

PLATE 7B

Central neural structures labeled

MESENCEPHALON

TECTUM

Mesencephalic roof plate

Tectal NEP

TEGMENTUM

Tegmental NEP

Mesencephalic floor plate

Tectal primordial plexiform layer

Brain surface (heavier line)

MESENCEPHALIC PROTOVENTRICLE (FUTURE AQUEDUCT)

Tegmental primordial plexiform layer

SPINAL CORD

Spinal germinal zones

Spinal floor plate
(ventral commissural GEP)

Ventral NEP

Intermediate NEP

Dorsal NEP

Spinal roof plate

Primordial white matter

SLIT-SHAPED CENTRAL CANAL

Arrows indicate the presumed *direction of neuron migration* from germinal sources.

Arrows indicate the regionally *expanding shoreline* of the protoventricle with increase in stockbuilding NEP cells.

ABBREVIATIONS:
GEP - Glioepithelium
NEP - Neuroepithelium

FONT KEY:
VENTRICULAR DIVISIONS - CAPITALS
Germinal zone - Helvetica bold
Transient structure - Times bold italic
Permanent structure - Times Roman or **Bold**

PLATE 8A

**CR 4.0 mm, GW3.2
M714, Frontal/Horizontal
Section 63**

**Peripheral neural
and non-neural
structures
labeled**

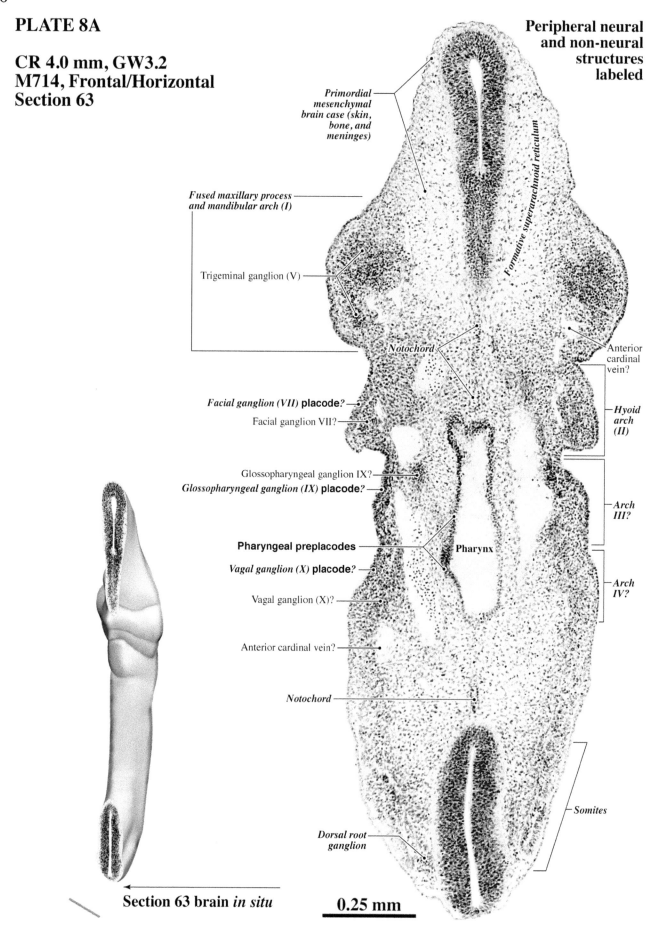

*Primordial
mesenchymal
brain case (skin,
bone, and
meninges)*

Formative superarachnoid reticulum

*Fused maxillary process
and mandibular arch (I)*

Trigeminal ganglion (V)

Notochord

Anterior
cardinal
vein?

Facial ganglion (VII) placode?

Facial ganglion VII?

*Hyoid
arch
(II)*

Glossopharyngeal ganglion IX?

Glossopharyngeal ganglion (IX) placode?

*Arch
III?*

Pharyngeal preplacodes

Pharynx

Vagal ganglion (X) placode?

Vagal ganglion (X)?

*Arch
IV?*

Anterior cardinal vein?

Notochord

Somites

*Dorsal root
ganglion*

← **Section 63 brain *in situ***

0.25 mm

Central neural structures labeled

MESENCEPHALON

TECTUM

Mesencephalic roof plate

Tectal NEP

TEGMENTUM/ISTHMUS

Tegmental NEP

Mesencephalic floor plate
(raphe glial system GEP)

Isthmal NEP

Metencephalic floor plate
(raphe glial system GEP)

PONS

RHOMBENCEPHALON

Tectal primordial plexiform layer

Brain surface (heavier line)

**MESENCEPHALIC PROTOVENTRICLE
(FUTURE AQUEDUCT)**

Tegmental primordial plexiform layer

Peripheral neural structures

Migrating trigeminal ganglionic neurons from the **trigeminal placode** *in the fused maxillary process and mandibular arch*

Migrating facial ganglionic neurons from the **facial placode** *in the hyoid arch*

Migrating glossopharyngeal ganglionic neurons from the **glossopharyngeal placode** *in arch III*

Migrating vagal ganglionic neurons from the **vagal placode** *in arch IV*

SPINAL CORD

Spinal germinal zones

Spinal floor plate
(ventral commissural GEP)

Ventral NEP

Intermediate NEP

Dorsal NEP

Spinal roof plate

Primordial white matter

SLIT-SHAPED CENTRAL CANAL

Arrows indicate the presumed *direction of neuron migration* from germinal sources.

Arrows indicate the regionally *expanding shoreline* of the protoventricle with increase in stockbuilding NEP cells.

**ABBREVIATIONS:
GEP - Glioepithelium
NEP - Neuroepithelium**

FONT KEY:
VENTRICULAR DIVISIONS - CAPITALS
Germinal zone - Helvetica bold
Transient structure - Times bold italic
Permanent structure - Times Roman or **Bold**

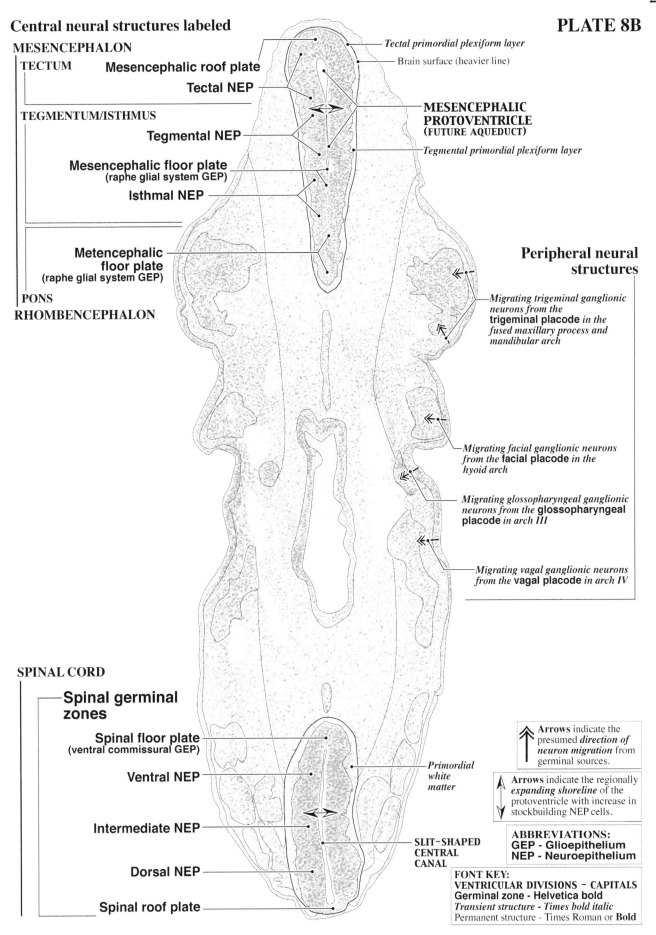

28

PLATE 9A

**CR 4.0 mm, GW3.2
M714, Frontal/Horizontal
Section 68**

**Peripheral neural
and non-neural
structures
labeled**

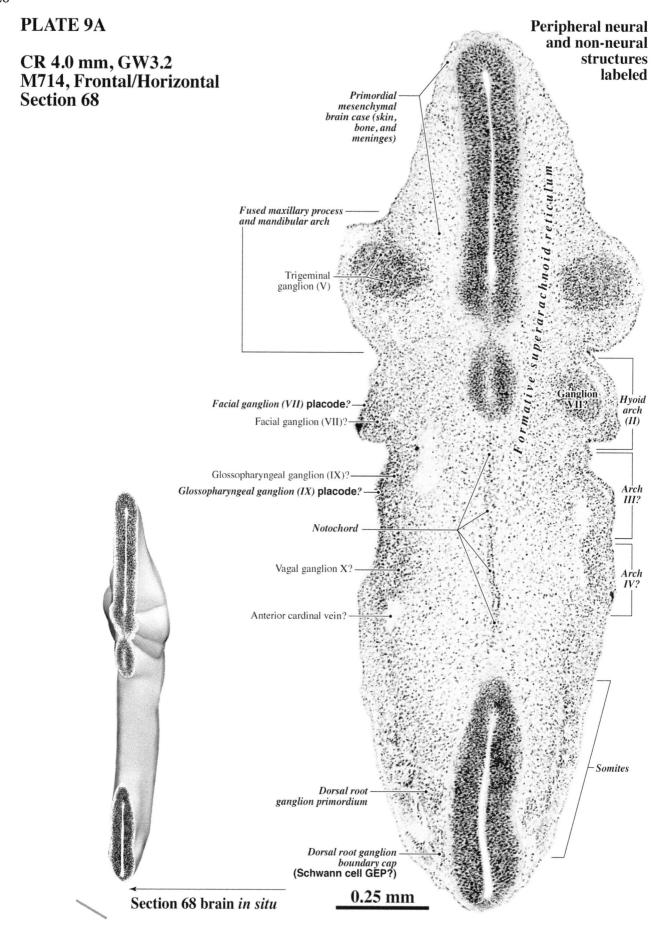

*Primordial
mesenchymal
brain case (skin,
bone, and
meninges)*

*Fused maxillary process
and mandibular arch*

Trigeminal
ganglion (V)

Formative superarachnoid reticulum

Ganglion
VII?

*Hyoid
arch
(II)*

Facial ganglion (VII) placode?

Facial ganglion (VII)?

*Arch
III?*

Glossopharyngeal ganglion (IX)?

Glossopharyngeal ganglion (IX) placode?

Notochord

*Arch
IV?*

Vagal ganglion X?

Anterior cardinal vein?

Somites

*Dorsal root
ganglion primordium*

*Dorsal root ganglion
boundary cap
(Schwann cell GEP?)*

← **Section 68 brain** *in situ*

0.25 mm

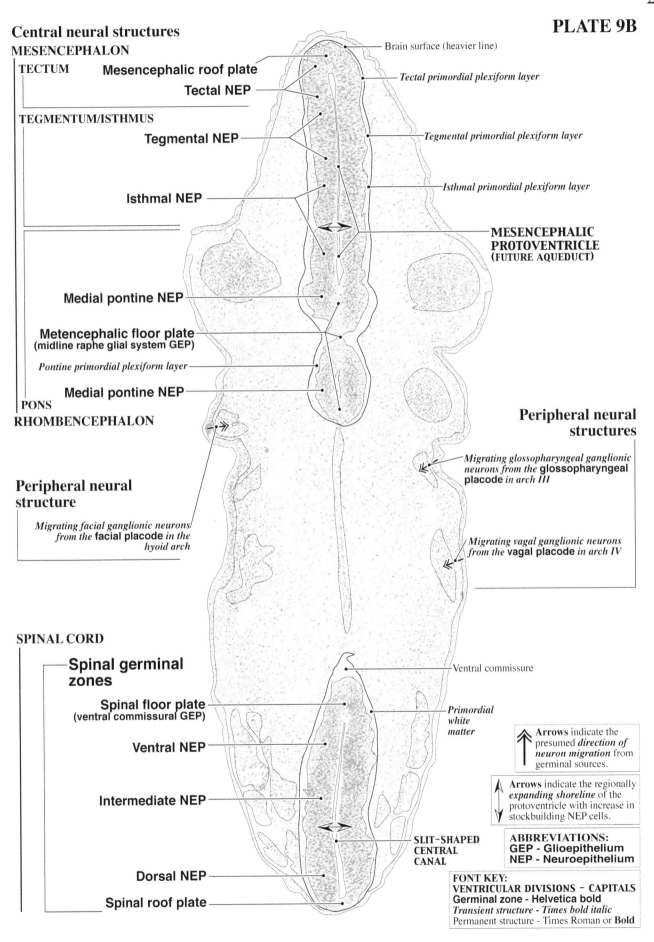

Central neural structures
MESENCEPHALON
TECTUM — Mesencephalic roof plate
Tectal NEP

TEGMENTUM/ISTHMUS
Tegmental NEP

Isthmal NEP

Medial pontine NEP

Metencephalic floor plate
(midline raphe glial system GEP)

Pontine primordial plexiform layer

Medial pontine NEP

PONS
RHOMBENCEPHALON

Peripheral neural structure
Migrating facial ganglionic neurons from the **facial placode** *in the hyoid arch*

SPINAL CORD

Spinal germinal zones

Spinal floor plate
(ventral commissural GEP)

Ventral NEP

Intermediate NEP

Dorsal NEP

Spinal roof plate

Brain surface (heavier line)

Tectal primordial plexiform layer

Tegmental primordial plexiform layer

Isthmal primordial plexiform layer

**MESENCEPHALIC PROTOVENTRICLE
(FUTURE AQUEDUCT)**

Peripheral neural structures

Migrating glossopharyngeal ganglionic neurons from the **glossopharyngeal placode** *in arch III*

Migrating vagal ganglionic neurons from the **vagal placode** *in arch IV*

Ventral commissure

Primordial white matter

SLIT-SHAPED CENTRAL CANAL

Arrows indicate the presumed *direction of neuron migration* from germinal sources.

Arrows indicate the regionally *expanding shoreline* of the protoventricle with increase in stockbuilding NEP cells.

ABBREVIATIONS:
GEP - Glioepithelium
NEP - Neuroepithelium

FONT KEY:
VENTRICULAR DIVISIONS - CAPITALS
Germinal zone - Helvetica bold
Transient structure - Times bold italic
Permanent structure - Times Roman or **Bold**

30

PLATE 10A

CR 4.0 mm, GW3.2
M714, Frontal/Horizontal
Section 73

**Peripheral neural and non-neural
structures labeled**

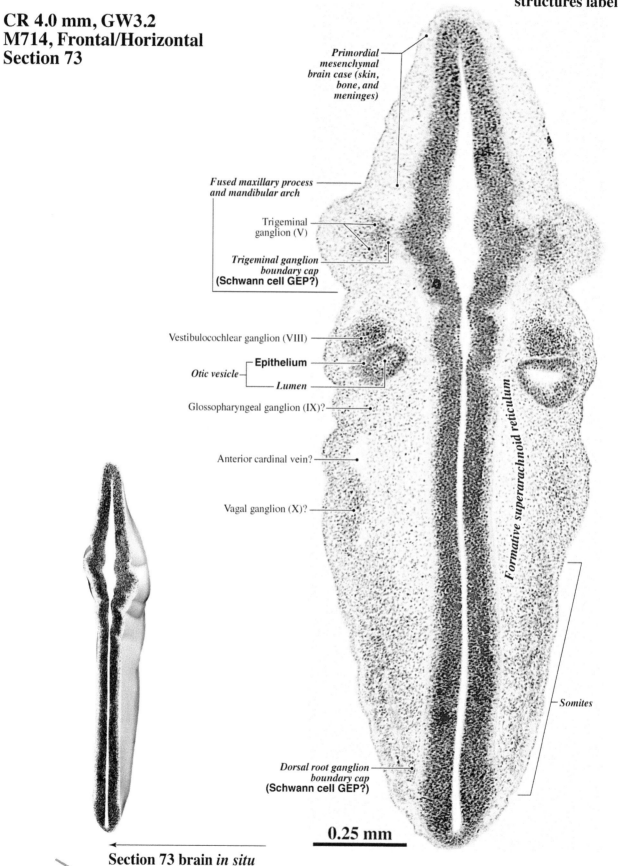

*Primordial
mesenchymal
brain case (skin,
bone, and
meninges)*

*Fused maxillary process
and mandibular arch*

Trigeminal
ganglion (V)

*Trigeminal ganglion
boundary cap*
(Schwann cell GEP?)

Vestibulocochlear ganglion (VIII)

Epithelium

Otic vesicle

Lumen

Glossopharyngeal ganglion (IX)?

Anterior cardinal vein?

Vagal ganglion (X)?

Formative superarachnoid reticulum

Somites

*Dorsal root ganglion
boundary cap*
(Schwann cell GEP?)

0.25 mm

Section 73 brain *in situ*

30

Central neural structures labeled

MESENCEPHALON
|ISTHMUS

Brain surface (heavier line)

Mesencephalic roof plate

Isthmal primordial plexiform layer

Isthmal NEP

MESENCEPHALIC PROTOVENTRICLE
(FUTURE AQUEDUCT)

|PONS/MEDULLA

RHOMBENCEPHALIC PROTOVENTRICLE
(FUTURE FOURTH VENTRICLE)

Cerebellar NEP?

R2

R3

R4+5

R6

R7

Medullary NEP
(vagal motor [X] and hypoglossal [XII]
NEPs blend with ventral spinal NEP)

RHOMBENCEPHALON
SPINAL CORD

Spinal germinal zones

Ventral NEP

Intermediate NEP

Dorsal NEP

Spinal roof plate

PROPOSED RHOMBOMERE IDENTITIES

R2 Trigeminal NEP - germinal source of the central trigeminal nuclei except the mesencephalic nucleus.

R3 Facial NEP - germinal source of central sensory receptor neurons getting input from the facial (VII) nerve.

R4 Vestibulo-auditory NEP - germinal source (with **R5**) of central auditory nuclei and vestibular nuclei, except the cochlear nuclei.

R5 Vestibulo-auditory NEP - germinal source (with **R4**) of central auditory nuclei and vestibular nuclei, except the cochlear nuclei.

R6 Glossopharyngeal NEP - germinal source of sensory neurons that receive input from the glossopharyngeal (IX) nerve.

R7 Vagal (X) sensory NEP - germinal source of the dorsal sensory nucleus and other sensory vagal nuclei.

Migrating vestibulocochlear ganglionic neurons from the otic epithelium

Peripheral neural structure

Primordial white matter

SLIT-SHAPED CENTRAL CANAL

ABBREVIATIONS:
GEP - Glioepithelium
NEP - Neuroepithelium
R - Rhombomere

Arrows indicate the presumed *direction of neuron migration* from germinal sources.

Arrows indicate the regionally *expanding shoreline* of the protoventricle with increase in stockbuilding NEP cells.

FONT KEY:
VENTRICULAR DIVISIONS - CAPITALS
Germinal zone - Helvetica bold
Transient structure - Times bold italic
Permanent structure - Times Roman or **Bold**

PLATE 11A

Peripheral neural and non-neural structures labeled

CR 4.0 mm, GW3.2
M714, Frontal/Horizontal
Section 78

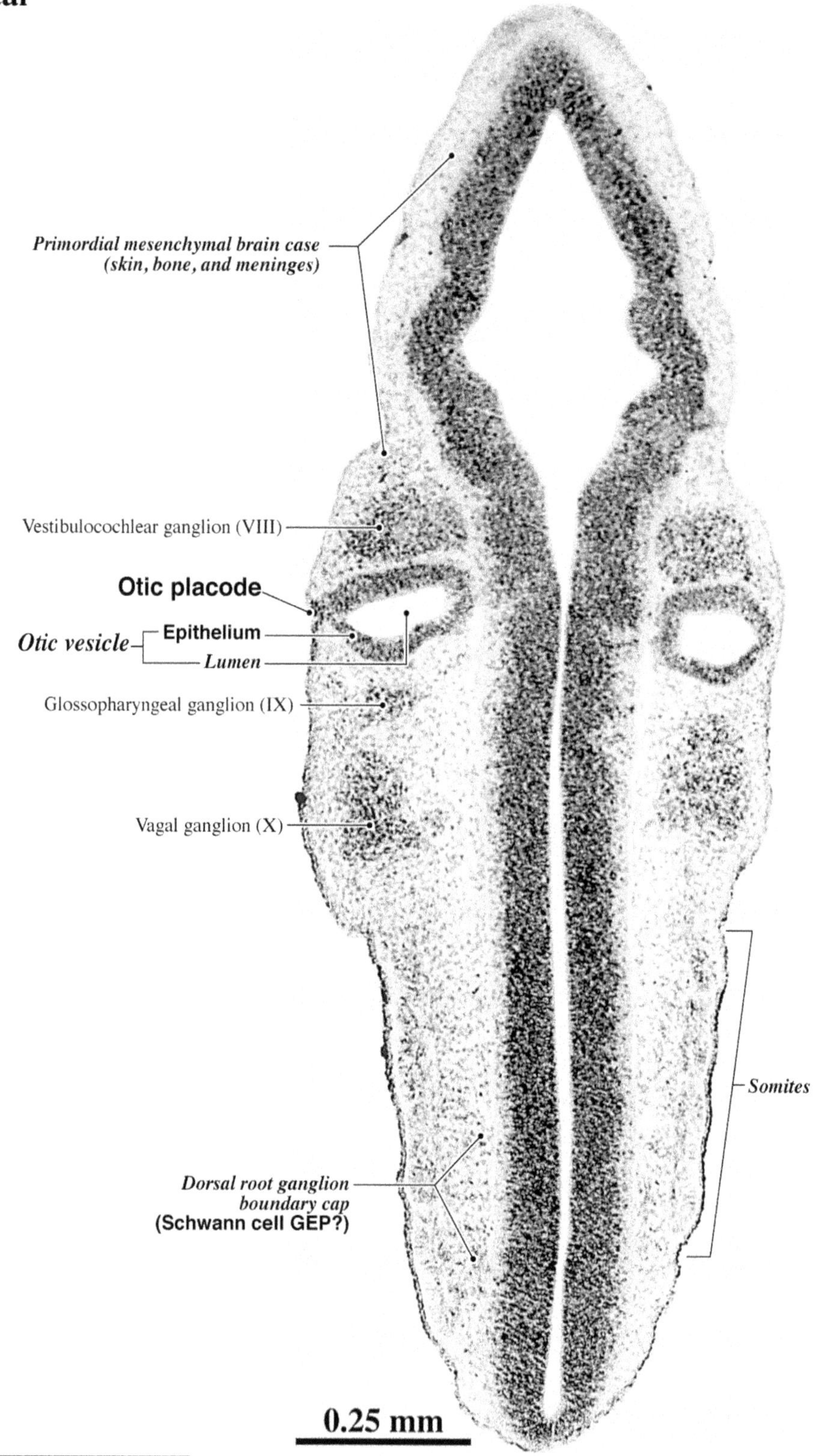

Primordial mesenchymal brain case
(skin, bone, and meninges)
Vestibulocochlear ganglion (VIII)
Otic placode
Otic vesicle
Epithelium
Lumen
Glossopharyngeal ganglion (IX)
Vagal ganglion (X)
Somites
Dorsal root ganglion
boundary cap
(Schwann cell GEP?)
0.25 mm

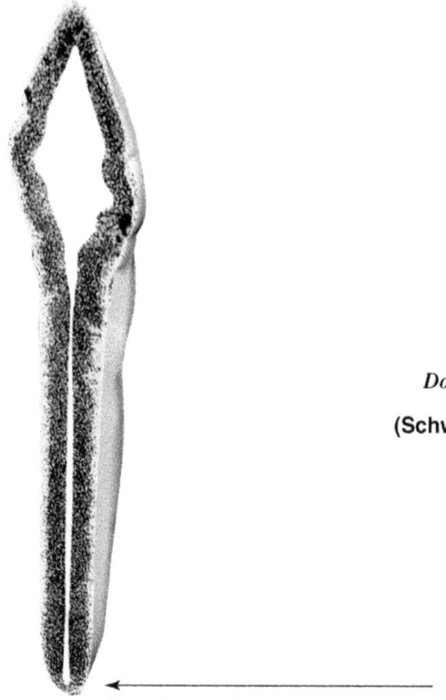

Section 78 brain *in situ*

Central neural structures labeled

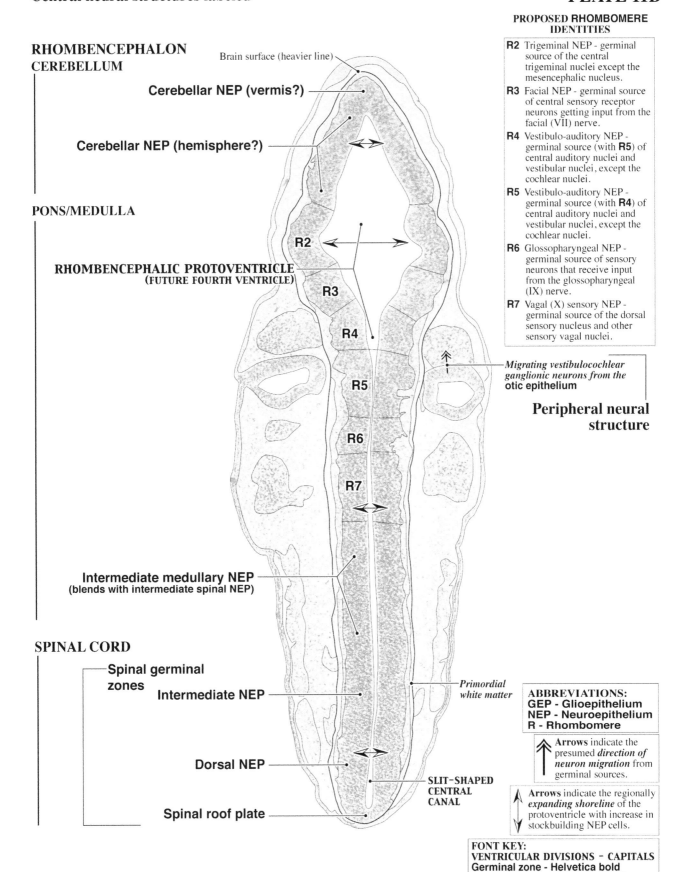

**RHOMBENCEPHALON
CEREBELLUM**

Brain surface (heavier line)

Cerebellar NEP (vermis?)

Cerebellar NEP (hemisphere?)

PONS/MEDULLA

R2

RHOMBENCEPHALIC PROTOVENTRICLE
(FUTURE FOURTH VENTRICLE)

R3

R4

R5

R6

R7

Intermediate medullary NEP
(blends with intermediate spinal NEP)

SPINAL CORD

Spinal germinal zones

Intermediate NEP

Dorsal NEP

Spinal roof plate

Primordial white matter

SLIT-SHAPED CENTRAL CANAL

PROPOSED RHOMBOMERE IDENTITIES

R2 Trigeminal NEP - germinal source of the central trigeminal nuclei except the mesencephalic nucleus.

R3 Facial NEP - germinal source of central sensory receptor neurons getting input from the facial (VII) nerve.

R4 Vestibulo-auditory NEP - germinal source (with **R5**) of central auditory nuclei and vestibular nuclei, except the cochlear nuclei.

R5 Vestibulo-auditory NEP - germinal source (with **R4**) of central auditory nuclei and vestibular nuclei, except the cochlear nuclei.

R6 Glossopharyngeal NEP - germinal source of sensory neurons that receive input from the glossopharyngeal (IX) nerve.

R7 Vagal (X) sensory NEP - germinal source of the dorsal sensory nucleus and other sensory vagal nuclei.

Migrating vestibulocochlear ganglionic neurons from the otic epithelium

Peripheral neural structure

ABBREVIATIONS:
GEP - Glioepithelium
NEP - Neuroepithelium
R - Rhombomere

Arrows indicate the presumed *direction of neuron migration* from germinal sources.

Arrows indicate the regionally *expanding shoreline* of the protoventricle with increase in stockbuilding NEP cells.

FONT KEY:
VENTRICULAR DIVISIONS - CAPITALS
Germinal zone - Helvetica bold
Transient structure - Times bold italic
Permanent structure - Times Roman or **Bold**

PLATE 12A

Peripheral neural and non-neural structures labeled

CR 4.0 mm, GW3.2
M714, Frontal/Horizontal
Section 83

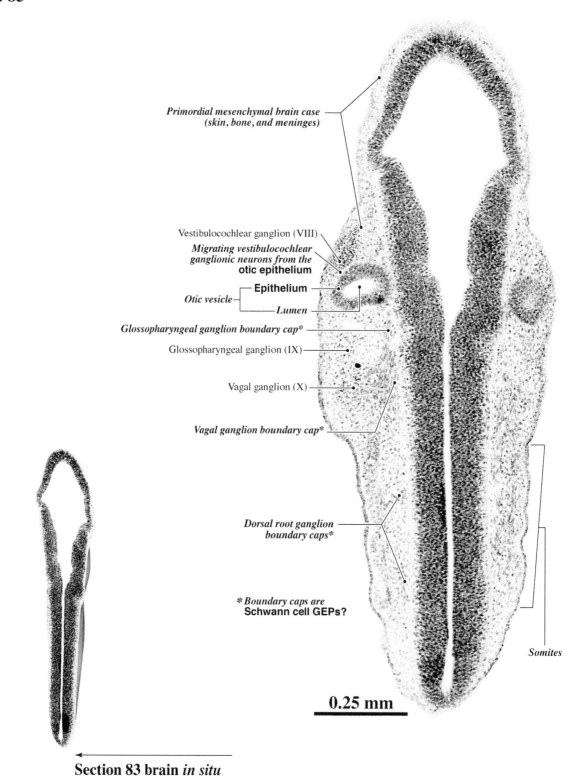

Primordial mesenchymal brain case
(skin, bone, and meninges)

Vestibulocochlear ganglion (VIII)

Migrating vestibulocochlear
ganglionic neurons from the
otic epithelium

Epithelium

Otic vesicle

Lumen

Glossopharyngeal ganglion boundary cap*

Glossopharyngeal ganglion (IX)

Vagal ganglion (X)

Vagal ganglion boundary cap*

Dorsal root ganglion
boundary caps*

* Boundary caps are
Schwann cell GEPs?

Somites

0.25 mm

Section 83 brain *in situ*

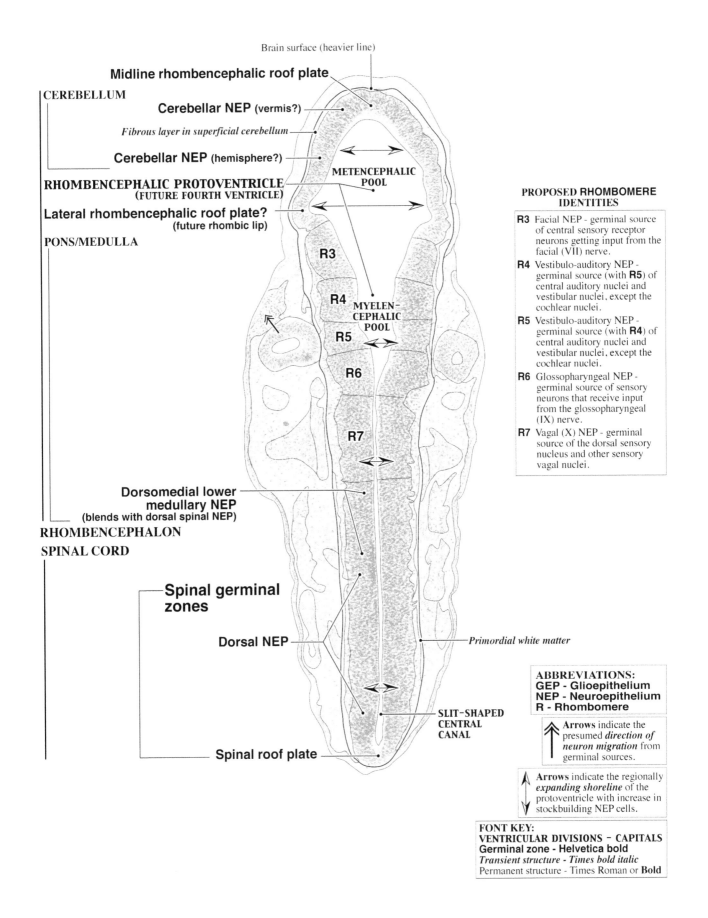

Brain surface (heavier line)

Midline rhombencephalic roof plate

CEREBELLUM

Cerebellar NEP (vermis?)

Fibrous layer in superficial cerebellum

Cerebellar NEP (hemisphere?)

RHOMBENCEPHALIC PROTOVENTRICLE
(FUTURE FOURTH VENTRICLE)

Lateral rhombencephalic roof plate?
(future rhombic lip)

PONS/MEDULLA

METENCEPHALIC POOL

R3

R4

MYELEN-CEPHALIC POOL

R5

R6

R7

Dorsomedial lower medullary NEP
(blends with dorsal spinal NEP)

RHOMBENCEPHALON

SPINAL CORD

Spinal germinal zones

Dorsal NEP

Primordial white matter

SLIT-SHAPED CENTRAL CANAL

Spinal roof plate

PROPOSED RHOMBOMERE IDENTITIES

R3 Facial NEP - germinal source of central sensory receptor neurons getting input from the facial (VII) nerve.

R4 Vestibulo-auditory NEP - germinal source (with **R5**) of central auditory nuclei and vestibular nuclei, except the cochlear nuclei.

R5 Vestibulo-auditory NEP - germinal source (with **R4**) of central auditory nuclei and vestibular nuclei, except the cochlear nuclei.

R6 Glossopharyngeal NEP - germinal source of sensory neurons that receive input from the glossopharyngeal (IX) nerve.

R7 Vagal (X) NEP - germinal source of the dorsal sensory nucleus and other sensory vagal nuclei.

ABBREVIATIONS:
GEP - Glioepithelium
NEP - Neuroepithelium
R - Rhombomere

Arrows indicate the presumed *direction of neuron migration* from germinal sources.

Arrows indicate the regionally *expanding shoreline* of the protoventricle with increase in stockbuilding NEP cells.

FONT KEY:
VENTRICULAR DIVISIONS - CAPITALS
Germinal zone - Helvetica bold
Transient structure - Times bold italic
Permanent structure - Times Roman or **Bold**

36

PLATE 13A

**CR 4.0 mm, GW3.2
M714, Frontal/Horizontal**

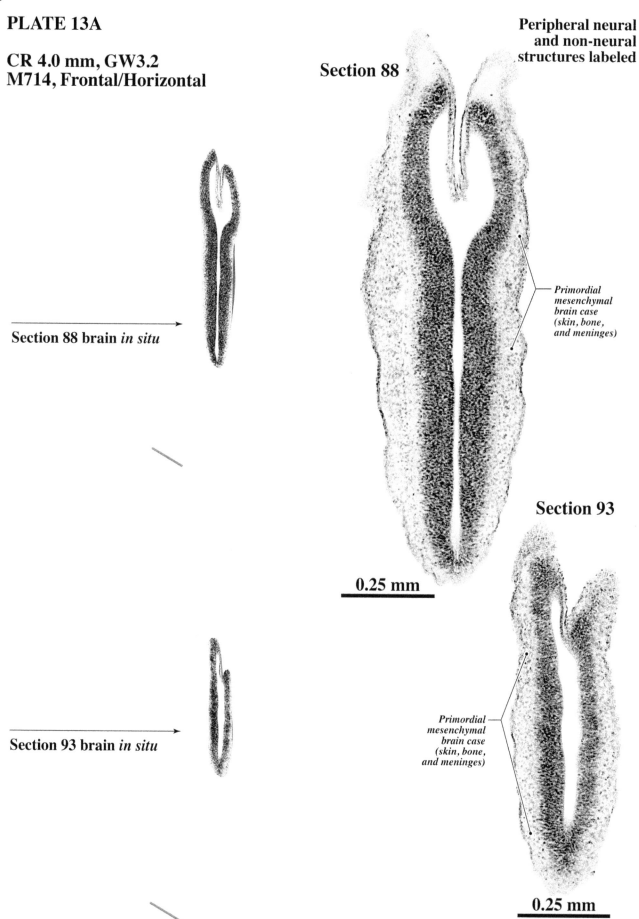

Section 88

Peripheral neural
and non-neural
structures labeled

Section 88 brain *in situ*

*Primordial
mesenchymal
brain case
(skin, bone,
and meninges)*

0.25 mm

Section 93

Section 93 brain *in situ*

*Primordial
mesenchymal
brain case
(skin, bone,
and meninges)*

0.25 mm

**Central neural
structures labeled**

Section 88

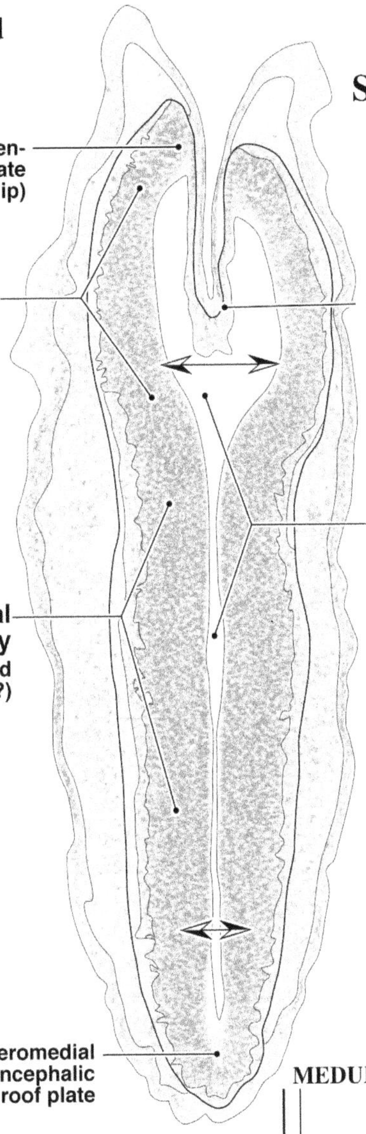

MEDULLA

Lateral myelen-
cephalic roof plate
(rhombic lip)

Medullary velum

**Future
precerebellar
and auditory**
(cochlear nuclear)
NEPs?

**RHOMBENCEPHALIC
PROTOVENTRICLE**
(FUTURE FOURTH VENTRICLE, MYELENCEPHALIC POOL)

**Dorsomedial
lower medullary
NEP** (gracile and
cuneate nuclei?)

Section 93

Posteromedial
myelencephalic
roof plate

MEDULLA

RHOMBENCEPHALON

Medullary velum

**Dorsomedial
lower medullary
NEP** (gracile and
cuneate nuclei?)

**RHOMBENCEPHALIC
PROTOVENTRICLE**
(FUTURE FOURTH
VENTRICLE, MYELEN-
CEPHALIC POOL)

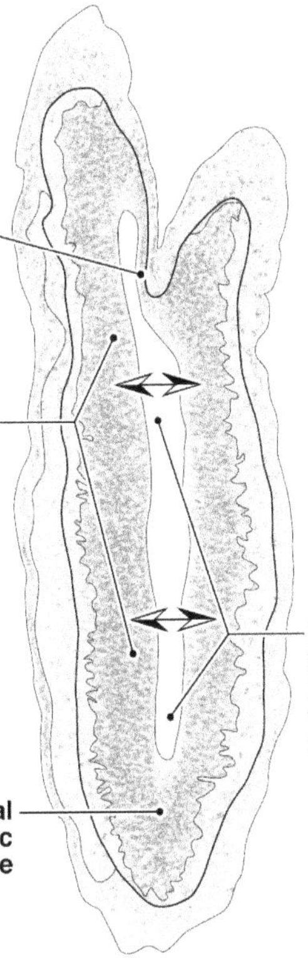

NEP - neuroepithelium

Arrows indicate the regionally
expanding shoreline of the
protoventricle with increase in
stockbuilding NEP cells.

FONT KEY:
VENTRICULAR DIVISIONS – CAPITALS
Germinal zone - Helvetica bold
Transient structure - Times bold italic
Permanent structure - Times Roman or **Bold**

Posteromedial
myelencephalic
roof plate

RHOMBENCEPHALON

PART III: C7724
CR 3.5 mm (GW 3.5)
Sagittal

Carnegie Collection specimen #7724 (designated here as C7724) has a 3.5-mm crown-rump length (CR). However, at this early stage, CR length is an unreliable estimate of gestational age. The right side of the body has clearly separated 24 to 25 somites with both anterior and posterior neuropores closed. Using the timetables in Patten (1953) and Hamilton et al. (1959), we estimate that C7724 is at gestational week (GW) 3.5. C7724 was fixed in formalin, was embedded in a celloidin/paraffin mix, and was cut in 8-μm sagittal sections that were stained with hematoxylin and eosin. Various orientations of the computer-aided 3-D reconstruction of C836's brain are used to show the gross external features of a GW3.5 brain (**Figure 7**). Like most sagittally cut specimens, C7724's sections are not parallel to the midline; **Figure 7** shows the approximate rotations in front (**B**) and back views (**C**). We photographed 24 sections at low magnification from the left to right sides of the body. Eight of the sections, mainly from the left side of the body, are illustrated in **Plates 14AB to 21AB**. Each illustrated section shows the entire embryo. Labels in **A Plates** (normal-contrast images) identify the approximate midline, non-neural structures, peripheral neural structures, and brain ventricular divisions; labels in **B Plates** (low-contrast images) identify central neural structures. **Plates 22AB** to **26AB** show high-magnification views of several parts of the brain.

The prosencephalon is the smallest major brain structure with little distinction between a future telencephalon and diencephalon. The entire prosencephalic neuroepithelium is rapidly stockbuilding its various populations of neuronal and glial stem cells surrounding a small prosencephalic protoventricle. The ventral and lateral prosencephalon is surrounded by cephalic preplacodes at the surface (for example, the anterolateral olfactory placode) that are continuous with those extending into the roof of the developing oral cavity (for example, Rathke's pouch).

The mesencephalon has stockbuilding pretectal and tectal neuroepithelia and a relatively short anteroposterior length. The stockbuilding tegmental and isthmal neuroepithelia form a distinctive arch between the mesencephalic and diencephalic flexures. These neuroepithelia surround a small mesencephalic protoventricle. There is a very thin subpial fiber band in the tegmentum and isthmus. The mesencephalon itself is distinguished by a pronounced arch, the mesencephalic flexure.

The rhombencephalon is the largest brain structure. Rhombomeres 2 through 7 form well-defined swellings in the lateral neuroepithelium (**Plate 18**). Most sensory cranial ganglia and the otic vesicle are located directly lateral to the rhombomeres with which they interact. The trigeminal ganglion (source of sensory V axons) appears in sections lateral to the last section that contains rhombomere 2. The vestibulocochlear ganglion (VIII afferents) and the otic vesicle are lateral to the last section that contains rhombomeres 4 and 5. The presumptive glossopharyngeal ganglion (IX afferents) is ventrolateral to the last section with rhombomere 6, and the presumptive vagal nerve (X afferents) is lateral to the last section with rhombomere 7. The presumptive facial ganglion (VII afferents) is near a branchial placode in the hyoid arch, slightly posterior and ventrolateral to rhombomere 3. Each rhombomere has a thin layer of pioneer migrating neurons that are only visible in most lateral sections, where the outer edges of the rhombomeric neuroepithelium are cut tangentially. Sections through the midline show a smoother neuroepithelium. Some migrating cells are outside the lower medullary neuroepithelium. The primordial white matter in the spinal cord extends into the lower medulla. The cerebellum stands out as the most immature and smallest rhombencephalic structure that blends with the isthmal neuroepithelium laterally and the presumptive tectal neuroepithelium medially.

EXTERNAL FEATURES OF THE GW3.5 BRAIN

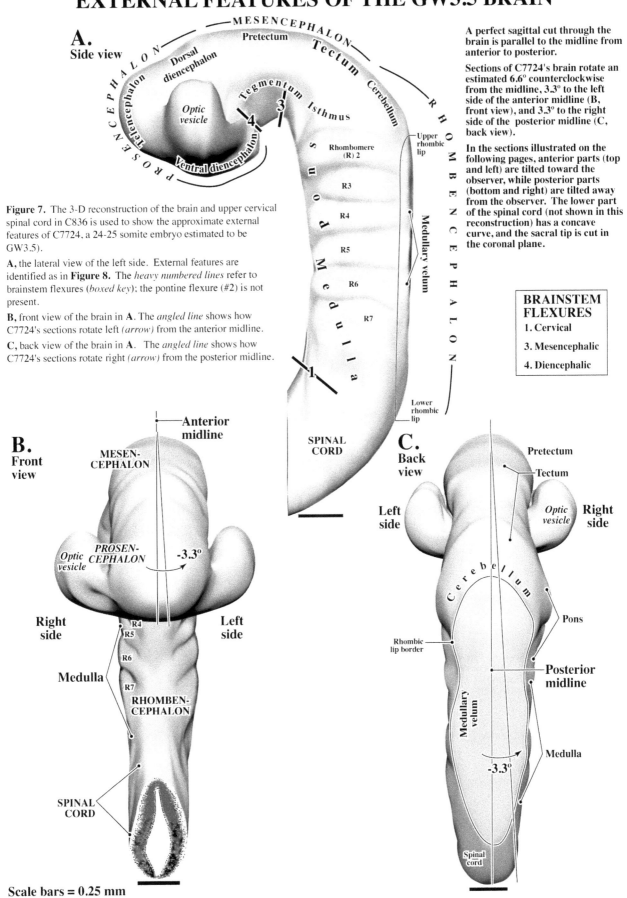

A. Side view

MESENCEPHALON
Pretectum
Tectum
Cerebellum
Isthmus
Tegmentum
PROSENCEPHALON
Dorsal diencephalon
Optic vesicle
Telencephalon
Ventral diencephalon
4
3
RHOMBENCEPHALON
Upper rhombic lip
Rhombomere (R) 2
R3
R4
R5
R6
R7
Medullary velum
Medulla
1
Lower rhombic lip
SPINAL CORD

A perfect sagittal cut through the brain is parallel to the midline from anterior to posterior.

Sections of C7724's brain rotate an estimated 6.6° counterclockwise from the midline, 3.3° to the left side of the anterior midline (B, front view), and 3.3° to the right side of the posterior midline (C, back view).

In the sections illustrated on the following pages, anterior parts (top and left) are tilted toward the observer, while posterior parts (bottom and right) are tilted away from the observer. The lower part of the spinal cord (not shown in this reconstruction) has a concave curve, and the sacral tip is cut in the coronal plane.

Figure 7. The 3-D reconstruction of the brain and upper cervical spinal cord in C836 is used to show the approximate external features of C7724, a 24-25 somite embryo estimated to be GW3.5).

A, the lateral view of the left side. External features are identified as in **Figure 8.** The *heavy numbered lines* refer to brainstem flexures (*boxed key*); the pontine flexure (#2) is not present.

B, front view of the brain in **A**. The *angled line* shows how C7724's sections rotate left (*arrow*) from the anterior midline.

C, back view of the brain in **A**. The *angled line* shows how C7724's sections rotate right (*arrow*) from the posterior midline.

BRAINSTEM FLEXURES
1. Cervical
3. Mesencephalic
4. Diencephalic

B. Front view

Anterior midline
MESEN-CEPHALON
PROSEN-CEPHALON
Optic vesicle
-3.3°
Right side
Left side
R4
R5
R6
R7
Medulla
RHOMBEN-CEPHALON
SPINAL CORD

C. Back view

Pretectum
Tectum
Optic vesicle
Left side
Right side
Cerebellum
Rhombic lip border
Pons
Posterior midline
Medullary velum
Medulla
-3.3°
Spinal cord

Scale bars = 0.25 mm

PLATE 14A

Medullary velum

RHOMBENCEPHALIC PROTOVENTRICLE
(FUTURE FOURTH VENTRICLE)

CR 3.5 mm, GW3.5
C7724, Sagittal
Slide 2
Section 30

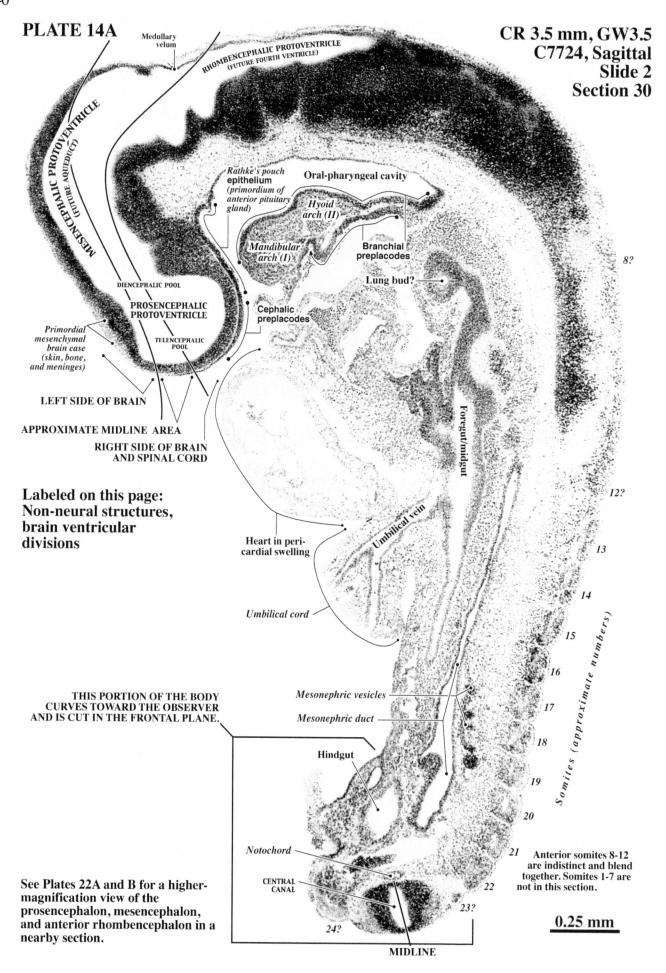

MESENCEPHALIC PROTOVENTRICLE
(FUTURE AQUEDUCT)

Rathke's pouch epithelium *(primordium of anterior pituitary gland)*

Oral-pharyngeal cavity

Hyoid arch (II)

Mandibular arch (I)

Branchial preplacodes

Lung bud?

8?

DIENCEPHALIC POOL

PROSENCEPHALIC PROTOVENTRICLE

Cephalic preplacodes

Primordial mesenchymal brain case (skin, bone, and meninges)

TELENCEPHALIC POOL

Foregut/midgut

LEFT SIDE OF BRAIN

APPROXIMATE MIDLINE AREA

RIGHT SIDE OF BRAIN AND SPINAL CORD

12?

Umbilical vein

13

Labeled on this page: Non-neural structures, brain ventricular divisions

Heart in peri-cardial swelling

14

15

Umbilical cord

16

Somites (approximate numbers)

THIS PORTION OF THE BODY CURVES TOWARD THE OBSERVER AND IS CUT IN THE FRONTAL PLANE.

Mesonephric vesicles

17

Mesonephric duct

18

19

Hindgut

20

21

Notochord

Anterior somites 8-12 are indistinct and blend together. Somites 1-7 are not in this section.

22

CENTRAL CANAL

23?

See Plates 22A and B for a higher-magnification view of the prosencephalon, mesencephalon, and anterior rhombencephalon in a nearby section.

24?

MIDLINE

0.25 mm

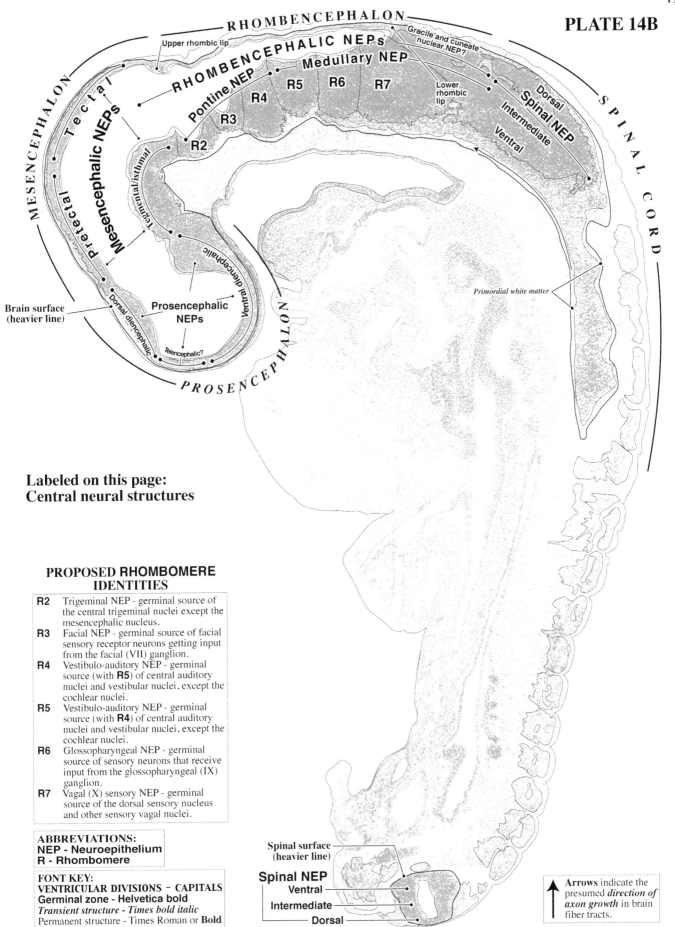

Labeled on this page:
Central neural structures

PROPOSED RHOMBOMERE IDENTITIES

R2	Trigeminal NEP - germinal source of the central trigeminal nuclei except the mesencephalic nucleus.
R3	Facial NEP - germinal source of facial sensory receptor neurons getting input from the facial (VII) ganglion.
R4	Vestibulo-auditory NEP - germinal source (with **R5**) of central auditory nuclei and vestibular nuclei, except the cochlear nuclei.
R5	Vestibulo-auditory NEP - germinal source (with **R4**) of central auditory nuclei and vestibular nuclei, except the cochlear nuclei.
R6	Glossopharyngeal NEP - germinal source of sensory neurons that receive input from the glossopharyngeal (IX) ganglion.
R7	Vagal (X) sensory NEP - germinal source of the dorsal sensory nucleus and other sensory vagal nuclei.

ABBREVIATIONS:
NEP - Neuroepithelium
R - Rhombomere

FONT KEY:
VENTRICULAR DIVISIONS – CAPITALS
Germinal zone - Helvetica bold
Transient structure - Times bold italic
Permanent structure - Times Roman or **Bold**

Arrows indicate the presumed *direction of axon growth* in brain fiber tracts.

42

PLATE 15A

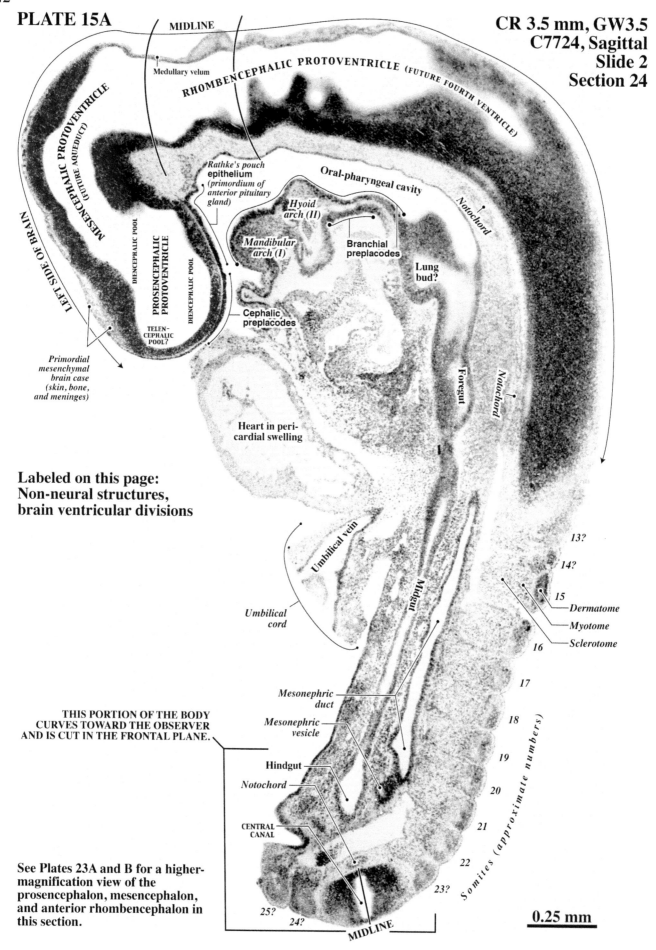

MIDLINE

Medullary velum

RHOMBENCEPHALIC PROTOVENTRICLE (FUTURE FOURTH VENTRICLE)

MESENCEPHALIC PROTOVENTRICLE (FUTURE AQUEDUCT)

Rathke's pouch epithelium *(primordium of anterior pituitary gland)*

Oral-pharyngeal cavity

Notochord

Hyoid arch (II)

Mandibular arch (I)

Branchial preplacodes

LEFT SIDE OF BRAIN

DIENCEPHALIC POOL

PROSENCEPHALIC PROTOVENTRICLE

DIENCEPHALIC POOL

Lung bud?

Cephalic preplacodes

TELEN-CEPHALIC POOL?

Primordial mesenchymal brain case (skin, bone, and meninges)

Foregut

Notochord

Heart in peri-cardial swelling

Labeled on this page: Non-neural structures, brain ventricular divisions

13?

14?

15

Umbilical vein

Dermatome

Myotome

Umbilical cord

16

Sclerotome

Midgut

17

Mesonephric duct

18

Mesonephric vesicle

THIS PORTION OF THE BODY CURVES TOWARD THE OBSERVER AND IS CUT IN THE FRONTAL PLANE.

19

Hindgut

20

Notochord

21

CENTRAL CANAL

22

23?

Somites (approximate numbers)

See Plates 23A and B for a higher-magnification view of the prosencephalon, mesencephalon, and anterior rhombencephalon in this section.

25?

24?

MIDLINE

0.25 mm

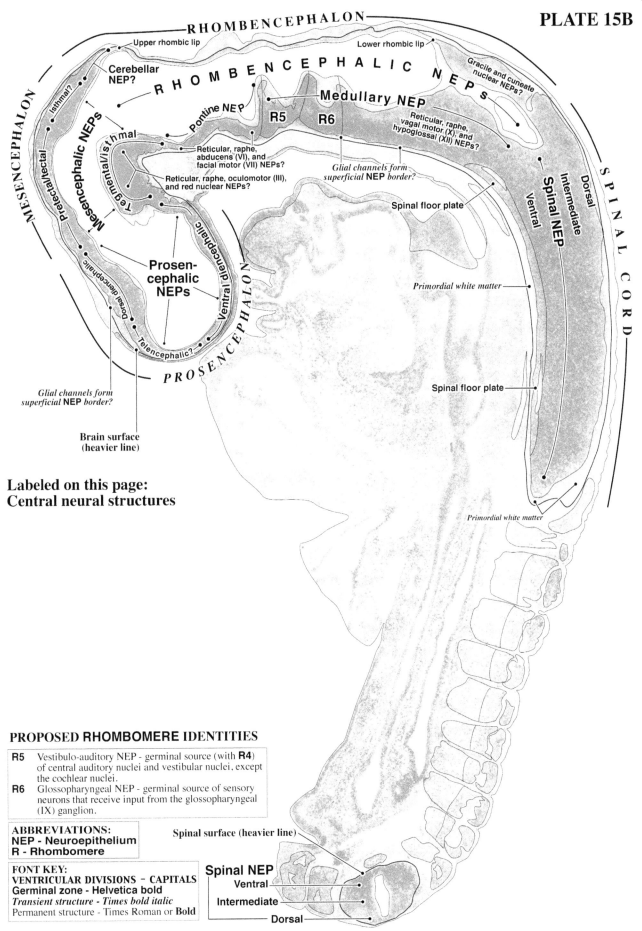

RHOMBENCEPHALON

Upper rhombic lip

Lower rhombic lip

Gracile and cuneate nuclear NEPs?

RHOMBENCEPHALIC NEPS

Cerebellar NEP?

Isthmal?

MESENCEPHALON

Pontine NEP

Medullary NEP

R5

R6

Reticular, raphe, vagal motor (X), and hypoglossal (XII) NEPs?

Mesencephalic NEPs

Tegmental/isthmal

Prefectal/tectal

Reticular, raphe, abducens (VI), and facial motor (VII) NEPs?

Reticular, raphe, oculomotor (III), and red nuclear NEPs?

Mesencephalic

Glial channels form superficial NEP border?

Spinal floor plate

Dorsal

Intermediate

Spinal NEP

Ventral

SPINAL CORD

Prosen-cephalic NEPs

Ventral diencephalic

Dorsal diencephalic

Primordial white matter

Telencephalic?

PROSENCEPHALON

Spinal floor plate

Glial channels form superficial NEP border?

Brain surface (heavier line)

Primordial white matter

**Labeled on this page:
Central neural structures**

PROPOSED RHOMBOMERE IDENTITIES

R5   Vestibulo-auditory NEP - germinal source (with **R4**) of central auditory nuclei and vestibular nuclei, except the cochlear nuclei.

R6   Glossopharyngeal NEP - germinal source of sensory neurons that receive input from the glossopharyngeal (IX) ganglion.

**ABBREVIATIONS:
NEP - Neuroepithelium
R - Rhombomere**

Spinal surface (heavier line)

**FONT KEY:
VENTRICULAR DIVISIONS - CAPITALS
Germinal zone - Helvetica bold
Transient structure - Times bold italic
Permanent structure - Times Roman or Bold**

Spinal NEP

Ventral

Intermediate

Dorsal

44

Primordial mesenchymal brain case (skin, bone, and meninges)

CR 3.5 mm, GW3.5
C7724, Sagittal
Slide 2
Section 20

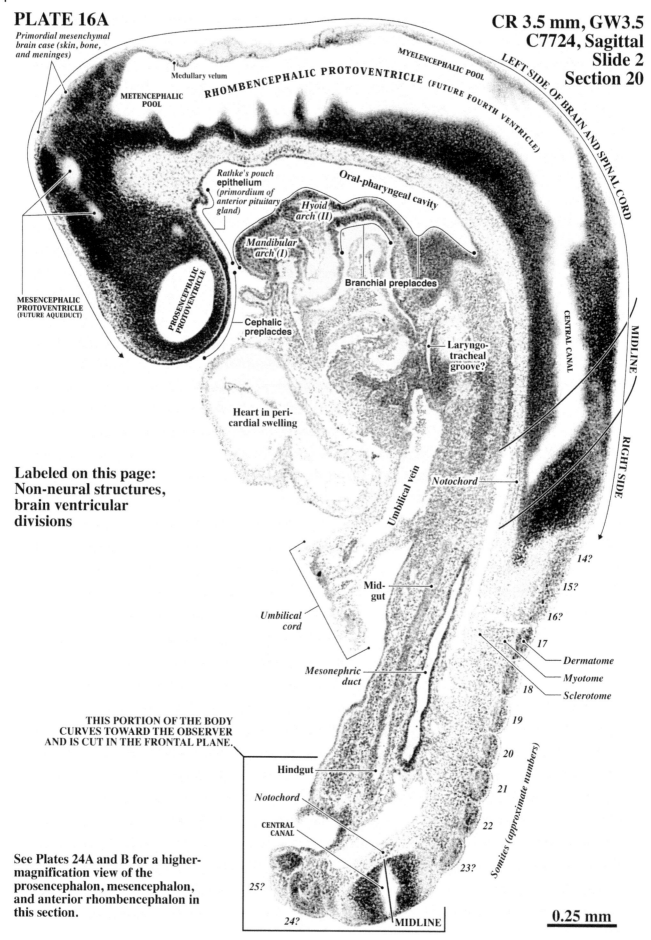

Medullary velum

METENCEPHALIC POOL

RHOMBENCEPHALIC PROTOVENTRICLE (FUTURE FOURTH VENTRICLE)

MYELENCEPHALIC POOL

LEFT SIDE OF BRAIN AND SPINAL CORD

Rathke's pouch epithelium (primordium of anterior pituitary gland)

Oral-pharyngeal cavity

Hyoid arch (II)

Mandibular arch (I)

Branchial preplacdes

MESENCEPHALIC PROTOVENTRICLE (FUTURE AQUEDUCT)

PROSENCEPHALIC PROTOVENTRICLE

Cephalic preplacdes

Laryngo-tracheal groove?

CENTRAL CANAL

MIDLINE

RIGHT SIDE

Heart in peri-cardial swelling

Labeled on this page: Non-neural structures, brain ventricular divisions

Umbilical vein

Notochord

14?

15?

16?

Mid-gut

17

Umbilical cord

Dermatome

Myotome

18

Sclerotome

Mesonephric duct

19

THIS PORTION OF THE BODY CURVES TOWARD THE OBSERVER AND IS CUT IN THE FRONTAL PLANE.

20

Somites (approximate numbers)

Hindgut

21

Notochord

22

CENTRAL CANAL

See Plates 24A and B for a higher-magnification view of the prosencephalon, mesencephalon, and anterior rhombencephalon in this section.

23?

25?

24?

MIDLINE

0.25 mm

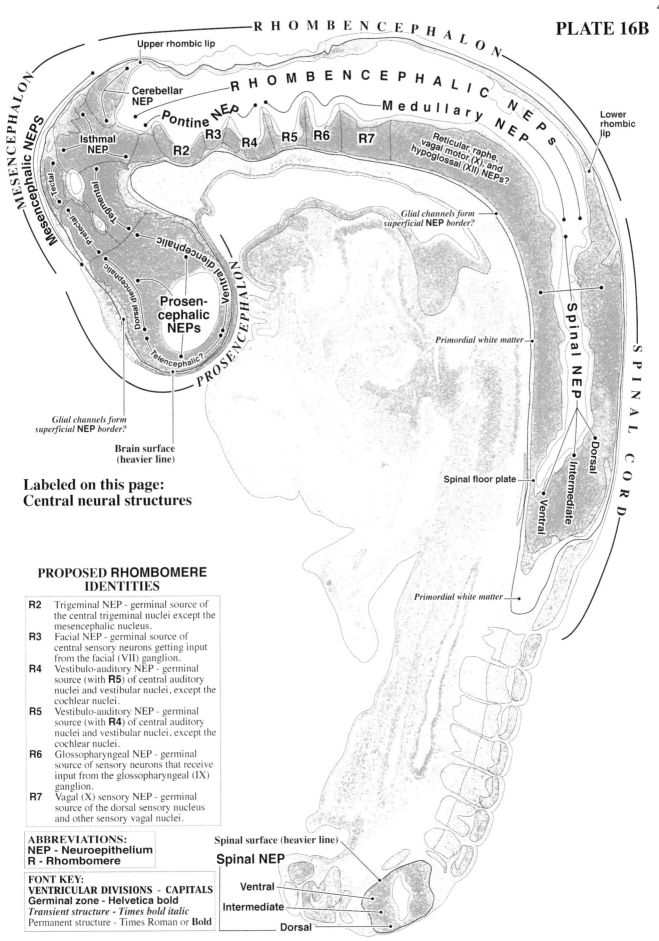

R H O M B E N C E P H A L O N

Upper rhombic lip

Cerebellar
NEP

R H O M B E N C E P H A L I C
N E P s

M e d u l l a r y   N E P

Lower
rhombic
lip

MESENCEPHALON

Mesencephalic NEPs

Isthmal
NEP

Pontine NEP

R2 R3 R4 R5 R6 R7

*Reticular, raphe,
vagal motor (X), and
hypoglossal (XII) NEPs?*

Tectal

Tegmental

Pretectal

Ventral diencephalic

Dorsal diencephalic

*Glial channels form
superficial NEP border?*

Spinal NEP

S P I N A L   C O R D

**Prosen-
cephalic
NEPs**

PROSENCEPHALON

Telencephalic?

Primordial white matter

Dorsal

Intermediate

Ventral

Spinal floor plate

*Glial channels form
superficial NEP border?*

**Brain surface
(heavier line)**

**Labeled on this page:
Central neural structures**

Primordial white matter

PROPOSED RHOMBOMERE IDENTITIES

R2 Trigeminal NEP - germinal source of the central trigeminal nuclei except the mesencephalic nucleus.

R3 Facial NEP - germinal source of central sensory neurons getting input from the facial (VII) ganglion.

R4 Vestibulo-auditory NEP - germinal source (with **R5**) of central auditory nuclei and vestibular nuclei, except the cochlear nuclei.

R5 Vestibulo-auditory NEP - germinal source (with **R4**) of central auditory nuclei and vestibular nuclei, except the cochlear nuclei.

R6 Glossopharyngeal NEP - germinal source of sensory neurons that receive input from the glossopharyngeal (IX) ganglion.

R7 Vagal (X) sensory NEP - germinal source of the dorsal sensory nucleus and other sensory vagal nuclei.

**ABBREVIATIONS:
NEP - Neuroepithelium
R - Rhombomere**

FONT KEY:
VENTRICULAR DIVISIONS - CAPITALS
Germinal zone - Helvetica bold
Transient structure - Times bold italic
Permanent structure - Times Roman or **Bold**

Spinal surface (heavier line)

Spinal NEP

Ventral

Intermediate

Dorsal

PLATE 17A

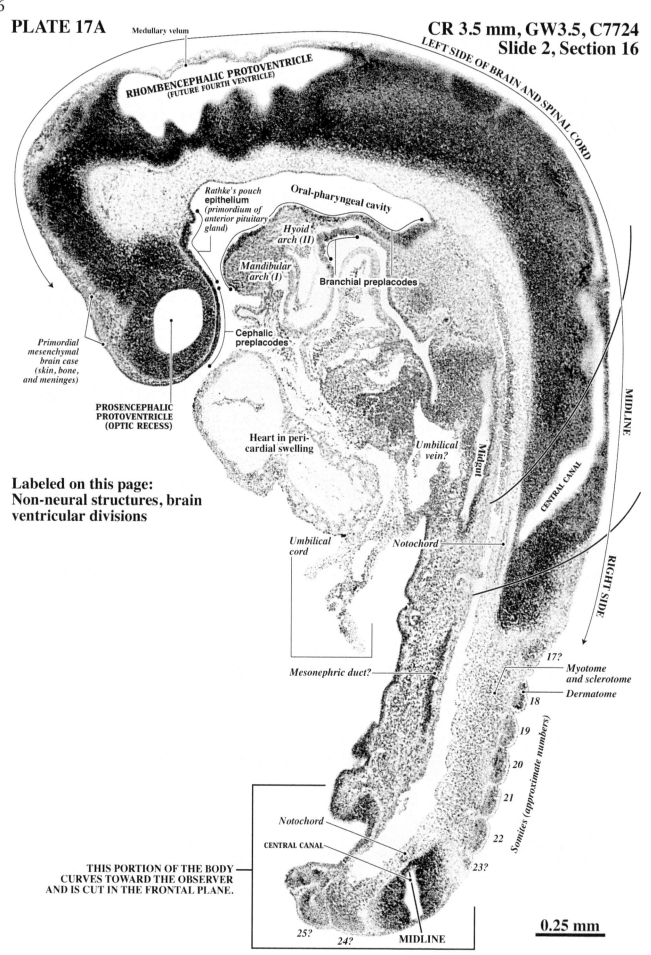

Medullary velum

RHOMBENCEPHALIC PROTOVENTRICLE
(FUTURE FOURTH VENTRICLE)

LEFT SIDE OF BRAIN AND SPINAL CORD

Rathke's pouch epithelium
(primordium of anterior pituitary gland)

Oral-pharyngeal cavity

Hyoid arch (II)

Mandibular arch (I)

Branchial preplacodes

Cephalic preplacodes

Primordial mesenchymal brain case (skin, bone, and meninges)

PROSENCEPHALIC PROTOVENTRICLE (OPTIC RECESS)

MIDLINE

CENTRAL CANAL

RIGHT SIDE

Labeled on this page: Non-neural structures, brain ventricular divisions

Heart in peri-cardial swelling

Umbilical vein?

Midgut

Umbilical cord

Notochord

Mesonephric duct?

17?

Myotome and sclerotome

Dermatome

18

19

20

21

Somites (approximate numbers)

22

Notochord

CENTRAL CANAL

23?

THIS PORTION OF THE BODY CURVES TOWARD THE OBSERVER AND IS CUT IN THE FRONTAL PLANE.

25?

24?

MIDLINE

0.25 mm

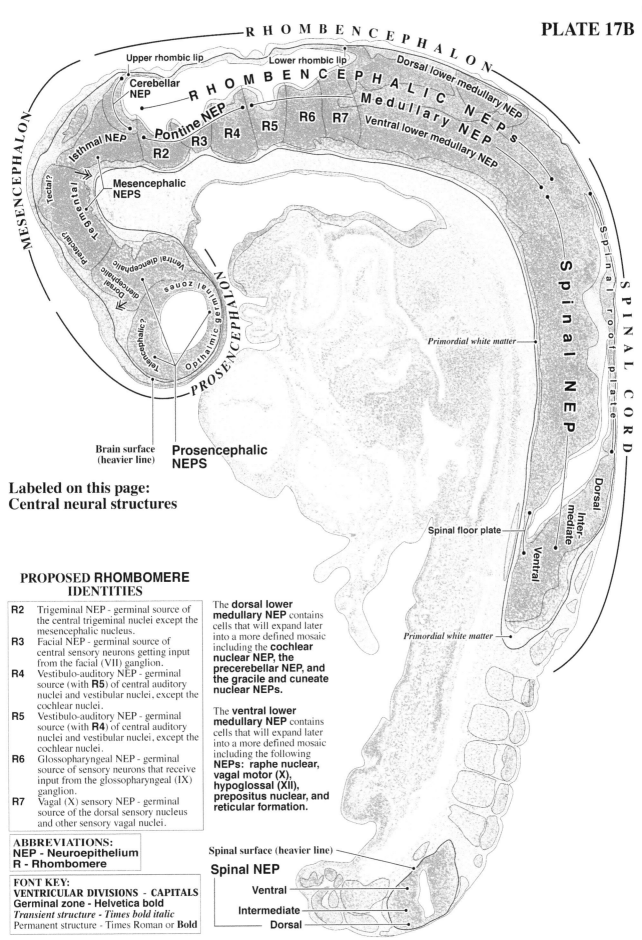

RHOMBENCEPHALON

Upper rhombic lip
Lower rhombic lip
Dorsal lower medullary NEP

Cerebellar NEP

RHOMBENCEPHALIC NEPs

Medullary NEP

Pontine NEP

R6 R7
R5
R4
R3
R2

Ventral lower medullary NEP

Isthmal NEP

MESENCEPHALON

Tectal?

Mesencephalic NEPS

Tegmental
pretectal

Dorsal diencephalic

Ventral diencephalic

Spinal roof plate

SPINAL CORD

Spinal NEP

Opthalmic germinal zones

germinal zones

Telencephalic?

PROSENCEPHALON

Primordial white matter

Brain surface (heavier line)

Prosencephalic NEPS

Dorsal
Inter-mediate
Ventral

Spinal floor plate

Primordial white matter

Labeled on this page: Central neural structures

PROPOSED RHOMBOMERE IDENTITIES

R2 Trigeminal NEP - germinal source of the central trigeminal nuclei except the mesencephalic nucleus.

R3 Facial NEP - germinal source of central sensory neurons getting input from the facial (VII) ganglion.

R4 Vestibulo-auditory NEP - germinal source (with **R5**) of central auditory nuclei and vestibular nuclei, except the cochlear nuclei.

R5 Vestibulo-auditory NEP - germinal source (with **R4**) of central auditory nuclei and vestibular nuclei, except the cochlear nuclei.

R6 Glossopharyngeal NEP - germinal source of sensory neurons that receive input from the glossopharyngeal (IX) ganglion.

R7 Vagal (X) sensory NEP - germinal source of the dorsal sensory nucleus and other sensory vagal nuclei.

ABBREVIATIONS:
NEP - Neuroepithelium
R - Rhombomere

FONT KEY:
VENTRICULAR DIVISIONS - CAPITALS
Germinal zone - Helvetica bold
Transient structure - Times bold italic
Permanent structure - Times Roman or **Bold**

The **dorsal lower medullary NEP** contains cells that will expand later into a more defined mosaic including the **cochlear nuclear NEP, the precerebellar NEP, and the gracile and cuneate nuclear NEPs.**

The **ventral lower medullary NEP** contains cells that will expand later into a more defined mosaic including the following **NEPs: raphe nuclear, vagal motor (X), hypoglossal (XII), prepositus nuclear, and reticular formation.**

Spinal surface (heavier line)

Spinal NEP

Ventral

Intermediate

Dorsal

48

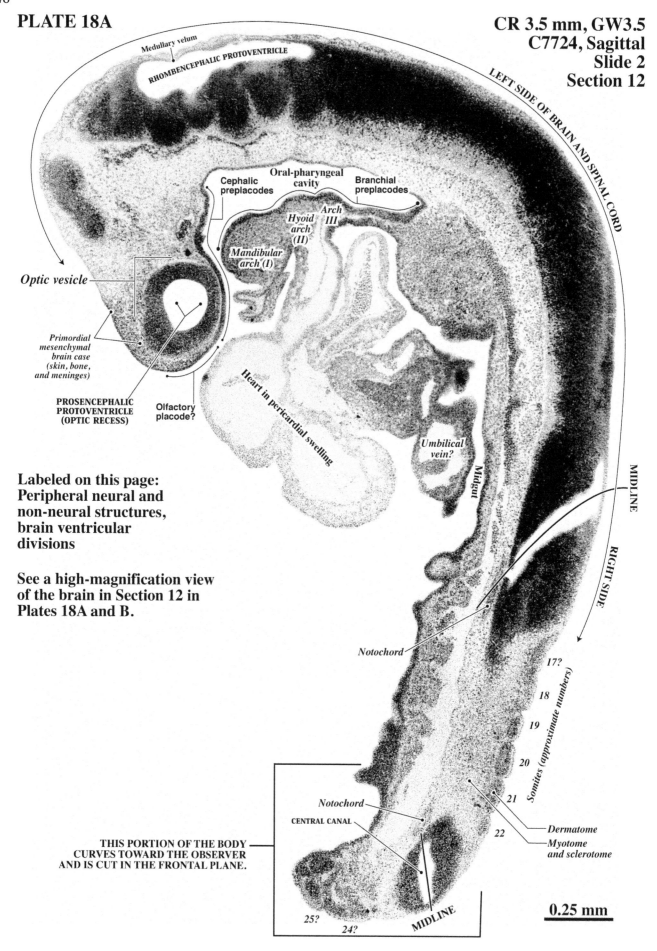

CR 3.5 mm, GW3.5
C7724, Sagittal
Slide 2
Section 12

LEFT SIDE OF BRAIN AND SPINAL CORD

Medullary velum

RHOMBENCEPHALIC PROTOVENTRICLE

Oral-pharyngeal
cavity

Cephalic
preplacodes

Branchial
preplacodes

Arch
III

Hyoid
arch
(II)

Mandibular
arch (I)

Optic vesicle

Primordial
mesenchymal
brain case
(skin, bone,
and meninges)

PROSENCEPHALIC
PROTOVENTRICLE
(OPTIC RECESS)

Olfactory
placode?

Heart in pericardial swelling

Umbilical
vein?

Midgut

MIDLINE

RIGHT SIDE

Labeled on this page:
Peripheral neural and
non-neural structures,
brain ventricular
divisions

See a high-magnification view
of the brain in Section 12 in
Plates 18A and B.

Notochord

17?

18

19

20

21

22

Somites (approximate numbers)

Notochord

CENTRAL CANAL

Dermatome

Myotome
and sclerotome

THIS PORTION OF THE BODY
CURVES TOWARD THE OBSERVER
AND IS CUT IN THE FRONTAL PLANE.

25?

24?

MIDLINE

0.25 mm

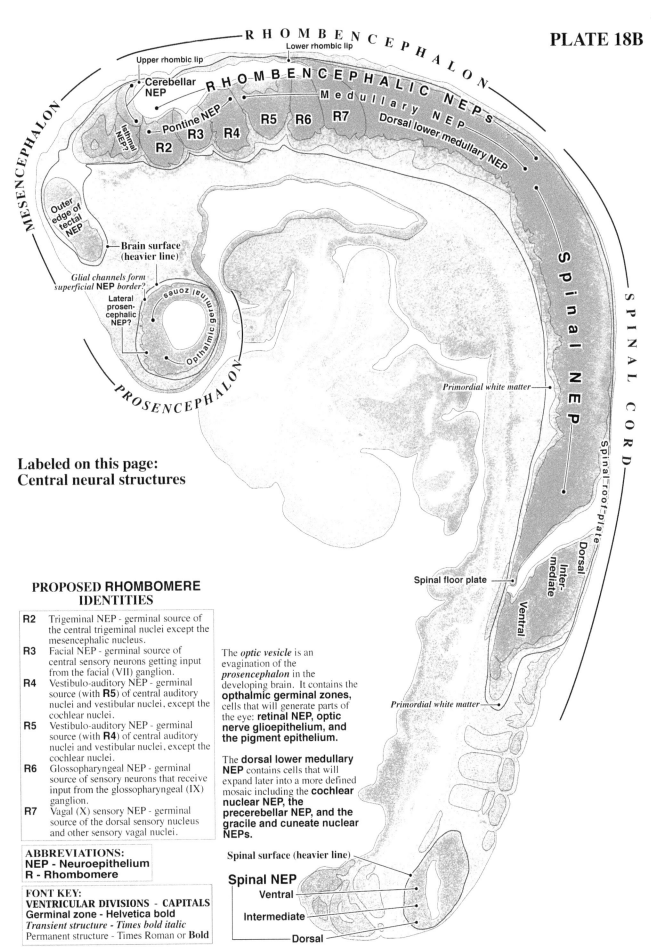

Labeled on this page:
Central neural structures

PROPOSED RHOMBOMERE IDENTITIES

R2 Trigeminal NEP - germinal source of the central trigeminal nuclei except the mesencephalic nucleus.

R3 Facial NEP - germinal source of central sensory neurons getting input from the facial (VII) ganglion.

R4 Vestibulo-auditory NEP - germinal source (with R5) of central auditory nuclei and vestibular nuclei, except the cochlear nuclei.

R5 Vestibulo-auditory NEP - germinal source (with R4) of central auditory nuclei and vestibular nuclei, except the cochlear nuclei.

R6 Glossopharyngeal NEP - germinal source of sensory neurons that receive input from the glossopharyngeal (IX) ganglion.

R7 Vagal (X) sensory NEP - germinal source of the dorsal sensory nucleus and other sensory vagal nuclei.

ABBREVIATIONS:
NEP - Neuroepithelium
R - Rhombomere

FONT KEY:
VENTRICULAR DIVISIONS - CAPITALS
Germinal zone - Helvetica bold
Transient structure - Times bold italic
Permanent structure - Times Roman or **Bold**

The *optic vesicle* is an evagination of the *prosencephalon* in the developing brain. It contains the **opthalmic germinal zones,** cells that will generate parts of the eye: **retinal NEP, optic nerve glioepithelium, and the pigment epithelium.**

The **dorsal lower medullary NEP** contains cells that will expand later into a more defined mosaic including the **cochlear nuclear NEP, the precerebellar NEP, and the gracile and cuneate nuclear NEPs.**

50

PLATE 19A

Medullary velum

RHOMBENCEPHALIC PROTOVENTRICLE
(FUTURE FOURTH VENTRICLE)

CR 3.5 mm
GW3.5
C7724
Sagittal
Slide 2
Section 8

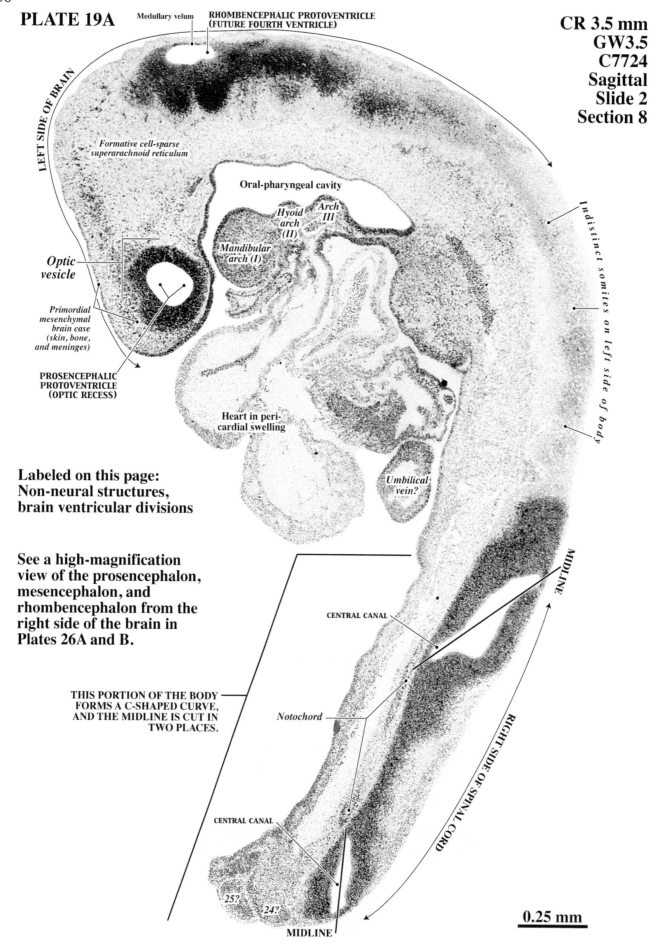

LEFT SIDE OF BRAIN

*Formative cell-sparse
superarachnoid reticulum*

Oral-pharyngeal cavity

*Hyoid
arch
(II)*

*Arch
III*

*Mandibular
arch (I)*

*Optic
vesicle*

*Primordial
mesenchymal
brain case
(skin, bone,
and meninges)*

**PROSENCEPHALIC
PROTOVENTRICLE
(OPTIC RECESS)**

Heart in peri-
cardial swelling

Umbilical
vein?

Indistinct somites on left side of body

**Labeled on this page:
Non-neural structures,
brain ventricular divisions**

**See a high-magnification
view of the prosencephalon,
mesencephalon, and
rhombencephalon from the
right side of the brain in
Plates 26A and B.**

MIDLINE

CENTRAL CANAL

**THIS PORTION OF THE BODY
FORMS A C-SHAPED CURVE,
AND THE MIDLINE IS CUT IN
TWO PLACES.**

Notochord

RIGHT SIDE OF SPINAL CORD

CENTRAL CANAL

25?

24?

MIDLINE

0.25 mm

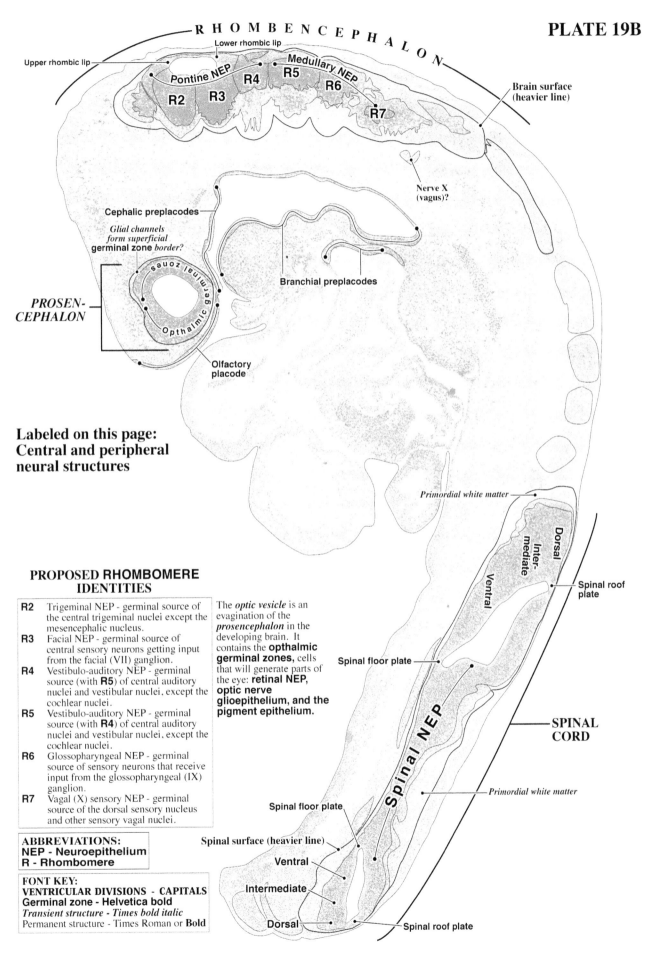

R H O M B E N C E P H A L O N

Upper rhombic lip
Lower rhombic lip
Pontine NEP
Medullary NEP
R2
R3
R4
R5
R6
R7
Brain surface
(heavier line)

Nerve X
(vagus)?

Cephalic preplacodes

*Glial channels
form superficial*
germinal zone *border?*

germinal zones

Ophthalmic

Branchial preplacodes

*PROSEN-
CEPHALON*

Olfactory
placode

**Labeled on this page:
Central and peripheral
neural structures**

Primordial white matter

Dorsal
Inter-
mediate
Ventral
Spinal roof
plate

PROPOSED RHOMBOMERE
IDENTITIES

R2	Trigeminal NEP - germinal source of the central trigeminal nuclei except the mesencephalic nucleus.
R3	Facial NEP - germinal source of central sensory neurons getting input from the facial (VII) ganglion.
R4	Vestibulo-auditory NEP - germinal source (with **R5**) of central auditory nuclei and vestibular nuclei, except the cochlear nuclei.
R5	Vestibulo-auditory NEP - germinal source (with **R4**) of central auditory nuclei and vestibular nuclei, except the cochlear nuclei.
R6	Glossopharyngeal NEP - germinal source of sensory neurons that receive input from the glossopharyngeal (IX) ganglion.
R7	Vagal (X) sensory NEP - germinal source of the dorsal sensory nucleus and other sensory vagal nuclei.

The *optic vesicle* is an evagination of the *prosencephalon* in the developing brain. It contains the **opthalmic germinal zones,** cells that will generate parts of the eye: **retinal NEP, optic nerve glioepithelium, and the pigment epithelium.**

Spinal floor plate

Spinal NEP

**SPINAL
CORD**

Primordial white matter

Spinal floor plate

Spinal surface (heavier line)

Ventral

Intermediate

Dorsal

Spinal roof plate

**ABBREVIATIONS:
NEP - Neuroepithelium
R - Rhombomere**

**FONT KEY:
VENTRICULAR DIVISIONS - CAPITALS
Germinal zone - Helvetica bold**
Transient structure - Times bold italic
Permanent structure - Times Roman or **Bold**

52

PLATE 20A

CR 3.5 mm
GW3.5
C7724
Sagittal
Slide 1
Section 38

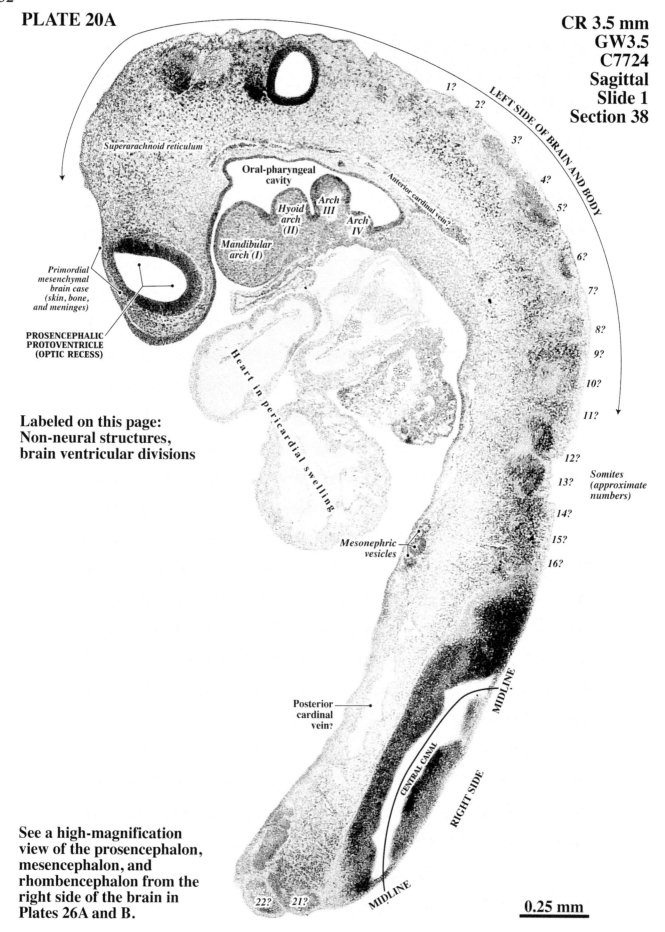

Superarachnoid reticulum

Oral-pharyngeal
cavity

LEFT SIDE OF BRAIN AND BODY

1?
2?
3?
4?
5?
6?
7?
8?
9?
10?
11?

Anterior cardinal vein?

*Arch
III*

*Hyoid
arch
(II)*

*Arch
IV*

*Mandibular
arch (I)*

*Primordial
mesenchymal
brain case
(skin, bone,
and meninges)*

PROSENCEPHALIC
PROTOVENTRICLE
(OPTIC RECESS)

Heart in pericardial swelling

Labeled on this page:
Non-neural structures,
brain ventricular divisions

12?
13?
14?
15?
16?

*Somites
(approximate
numbers)*

*Mesonephric
vesicles*

Posterior
cardinal
vein?

MIDLINE

CENTRAL CANAL

RIGHT SIDE

MIDLINE

See a high-magnification
view of the prosencephalon,
mesencephalon, and
rhombencephalon from the
right side of the brain in
Plates 26A and B.

22?   21?

0.25 mm

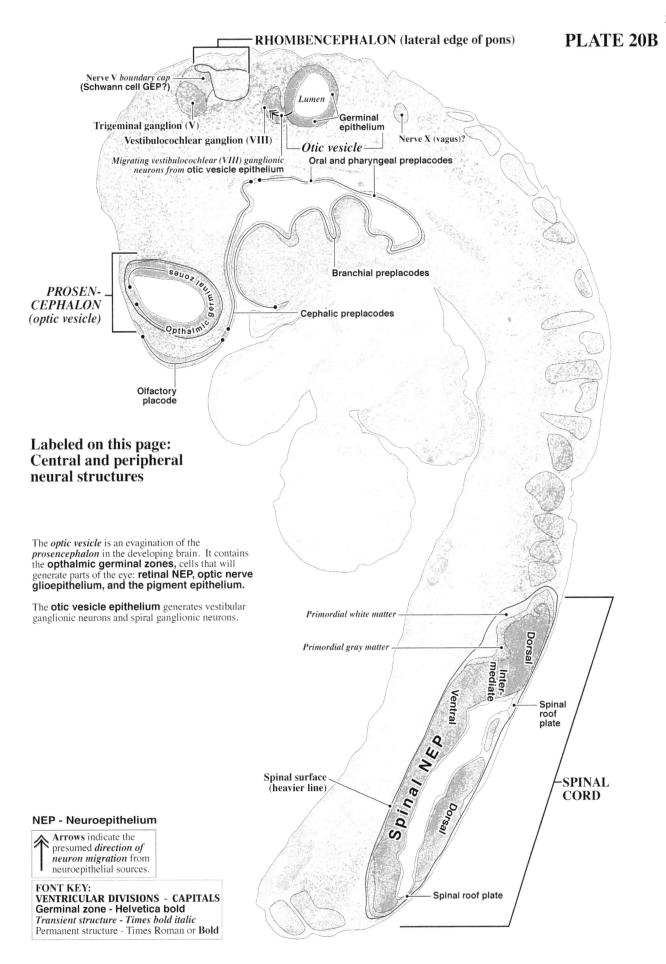

RHOMBENCEPHALON (lateral edge of pons)

PLATE 20B

Nerve V *boundary cap*
(Schwann cell GEP?)

Lumen

Trigeminal ganglion (V)

Germinal
epithelium

Vestibulocochlear ganglion (VIII)

Otic vesicle

Nerve X (vagus)?

*Migrating vestibulocochlear (VIII) ganglionic
neurons from* otic vesicle epithelium

Oral and pharyngeal preplacodes

*PROSEN-
CEPHALON
(optic vesicle)*

Opthalmic germinal zones

Branchial preplacodes

Cephalic preplacodes

Olfactory
placode

Labeled on this page:
Central and peripheral
neural structures

The *optic vesicle* is an evagination of the
prosencephalon in the developing brain. It contains
the **opthalmic germinal zones,** cells that will
generate parts of the eye: **retinal NEP, optic nerve
glioepithelium, and the pigment epithelium.**

The **otic vesicle epithelium** generates vestibular
ganglionic neurons and spiral ganglionic neurons.

Primordial white matter

Primordial gray matter

Dorsal

Inter-
mediate

Ventral

Spinal
roof
plate

Spinal NEP

Spinal surface
(heavier line)

Dorsal

**SPINAL
CORD**

Spinal roof plate

NEP - Neuroepithelium

Arrows indicate the
presumed *direction of
neuron migration* from
neuroepithelial sources.

FONT KEY:
VENTRICULAR DIVISIONS - CAPITALS
Germinal zone - Helvetica bold
Transient structure - Times bold italic
Permanent structure - Times Roman or **Bold**

54

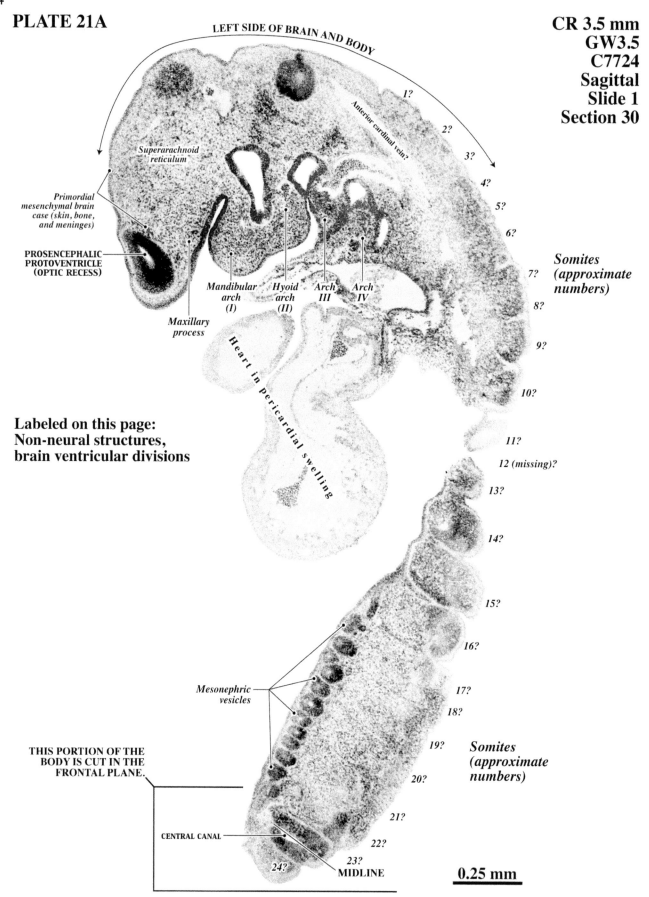

CR 3.5 mm
GW3.5
C7724
Sagittal
Slide 1
Section 30

LEFT SIDE OF BRAIN AND BODY

Anterior cardinal vein?

Superarachnoid reticulum

Primordial mesenchymal brain case (skin, bone, and meninges)

PROSENCEPHALIC PROTOVENTRICLE (OPTIC RECESS)

Mandibular arch (I)

Maxillary process

Hyoid arch (II)

Arch III

Arch IV

Heart in pericardial swelling

1?
2?
3?
4?
5?
6?

Somites (approximate numbers)

7?
8?
9?
10?
11?
12 (missing)?
13?
14?
15?
16?
17?
18?

Labeled on this page:
Non-neural structures,
brain ventricular divisions

Mesonephric vesicles

19?
20?
21?
22?

Somites (approximate numbers)

THIS PORTION OF THE BODY IS CUT IN THE FRONTAL PLANE.

CENTRAL CANAL

24?

23?
MIDLINE

0.25 mm

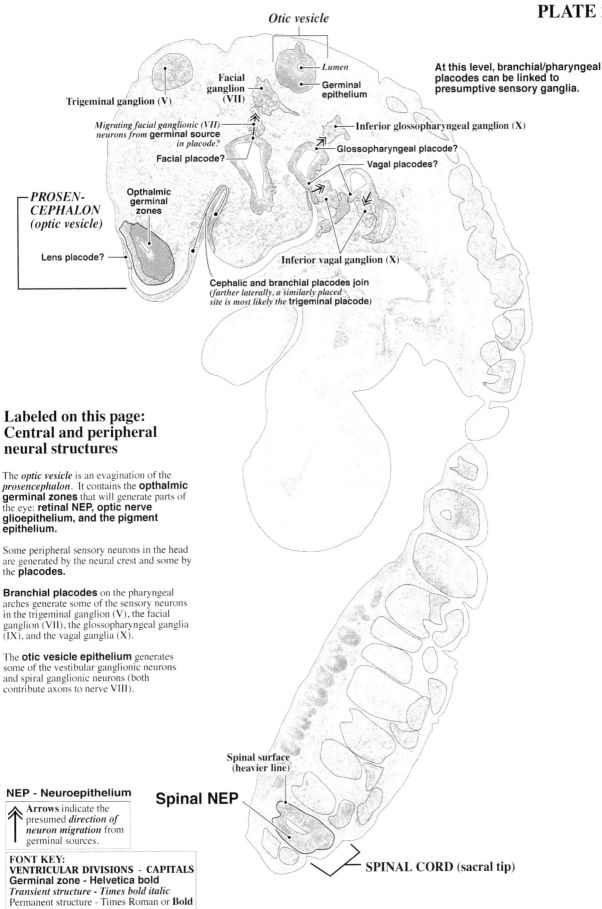

Otic vesicle

Lumen

Germinal
epithelium

**Facial
ganglion
(VII)**

Trigeminal ganglion (V)

At this level, branchial/pharyngeal
placodes can be linked to
presumptive sensory ganglia.

*Migrating facial ganglionic (VII)
neurons from* **germinal source**
in placode?

Facial placode?

Inferior glossopharyngeal ganglion (X)

Glossopharyngeal placode?

Vagal placodes?

*PROSEN-
CEPHALON
(optic vesicle)*

**Opthalmic
germinal
zones**

Lens placode?

Inferior vagal ganglion (X)

Cephalic and branchial placodes join
*(farther laterally, a similarly placed
site is most likely the* **trigeminal placode***)*

Labeled on this page:
Central and peripheral
neural structures

The *optic vesicle* is an evagination of the
prosencephalon. It contains the **opthalmic
germinal zones** that will generate parts of
the eye: **retinal NEP, optic nerve
glioepithelium, and the pigment
epithelium.**

Some peripheral sensory neurons in the head
are generated by the neural crest and some by
the **placodes.**

Branchial placodes on the pharyngeal
arches generate some of the sensory neurons
in the trigeminal ganglion (V), the facial
ganglion (VII), the glossopharyngeal ganglia
(IX), and the vagal ganglia (X).

The **otic vesicle epithelium** generates
some of the vestibular ganglionic neurons
and spiral ganglionic neurons (both
contribute axons to nerve VIII).

Spinal surface
(heavier line)

NEP - Neuroepithelium

Arrows indicate the
presumed *direction of
neuron migration* from
germinal sources.

Spinal NEP

FONT KEY:
VENTRICULAR DIVISIONS - CAPITALS
Germinal zone - Helvetica bold
Transient structure - Times bold italic
Permanent structure - Times Roman or **Bold**

SPINAL CORD (sacral tip)

56

CR 3.5 mm
GW3.5
C7724
Sagittal
Slide 2
Section 33

PROSENCEPHALON, MESENCEPHALON,
AND ANTERIOR RHOMBENCEPHALON

PLATE 22A
This section is from
the right side
of the brain.

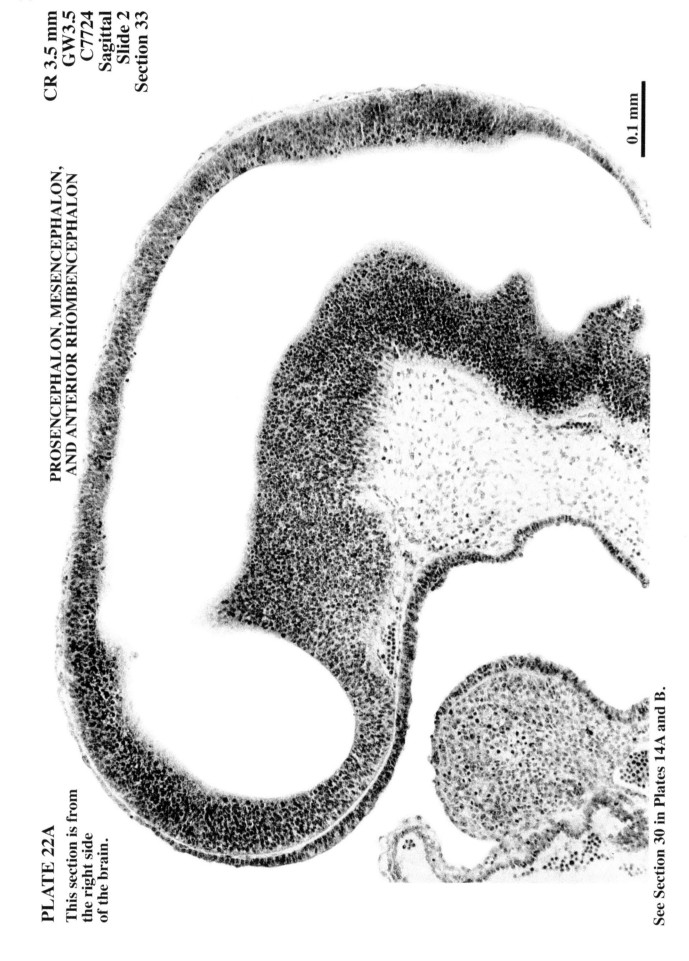

0.1 mm

See Section 30 in Plates 14A and B.

57

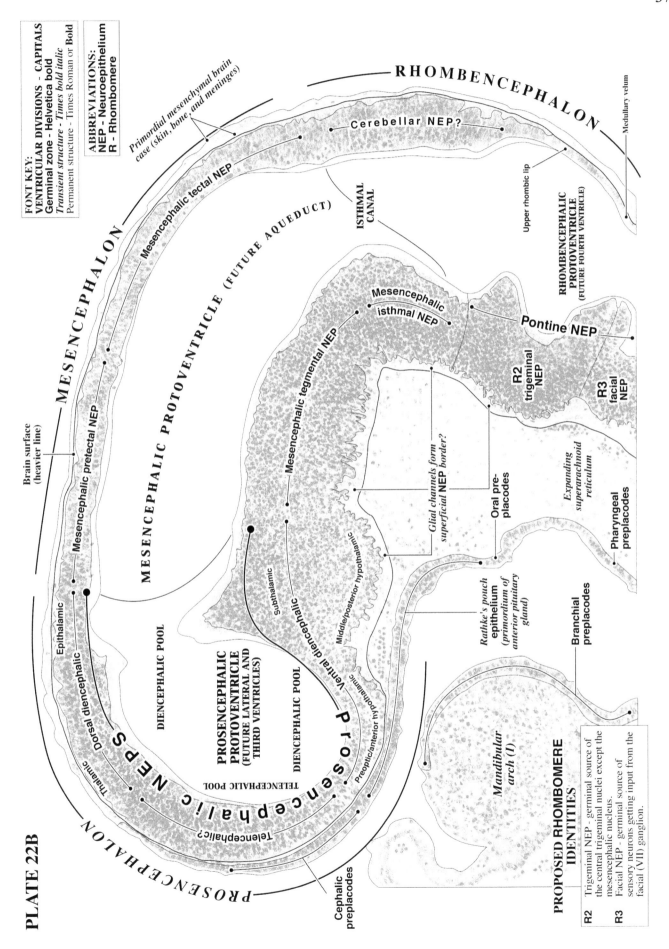

PLATE 22B

FONT KEY:
VENTRICULAR DIVISIONS - CAPITALS
Germinal zone - Helvetica bold
Transient structure - Times bold italic
Permanent structure - Times Roman or **Bold**

ABBREVIATIONS:
NEP - Neuroepithelium
R - Rhombomere

RHOMBENCEPHALON

MESENCEPHALON

PROSENCEPHALON

Primordial mesenchymal brain case (skin, bone, and meninges)

Cerebellar NEP?

Medullary velum

Mesencephalic tectal NEP

ISTHMAL CANAL

Upper rhombic lip

RHOMBENCEPHALIC PROTOVENTRICLE
(FUTURE FOURTH VENTRICLE)

MESENCEPHALIC PROTOVENTRICLE (FUTURE AQUEDUCT)

Mesencephalic isthmal NEP

Pontine NEP

Mesencephalic tegmental NEP

R2 trigeminal NEP

R3 facial NEP

Brain surface (heavier line)

Mesencephalic pretectal NEP

Epithalamic

MESENCEPHALIC PROTOVENTRICLE

Glial channels form superficial NEP border?

Expanding superarachnoid reticulum

Oral pre-placodes

Subthalamic

Middle/posterior hypothalamic

Pharyngeal preplacodes

DIENCEPHALIC POOL

PROSENCEPHALIC PROTOVENTRICLE
(FUTURE LATERAL AND THIRD VENTRICLES)

DIENCEPHALIC POOL

Ventral diencephalic

Rathke's pouch epithelium (primordium of anterior pituitary gland)

Preoptic/anterior hypothalamic

Branchial preplacodes

Dorsal diencephalic

TELENCEPHALIC POOL

Prosencephalic NEPS

Thalamic

Telencephalic?

Mandibular arch (I)

Cephalic preplacodes

PROSENCEPHALIC NEPS

PROPOSED RHOMBOMERE IDENTITIES

R2   Trigeminal NEP - germinal source of the central trigeminal nuclei except the mesencephalic nucleus.

R3   Facial NEP - germinal source of sensory neurons getting input from the facial (VII) ganglion.

58

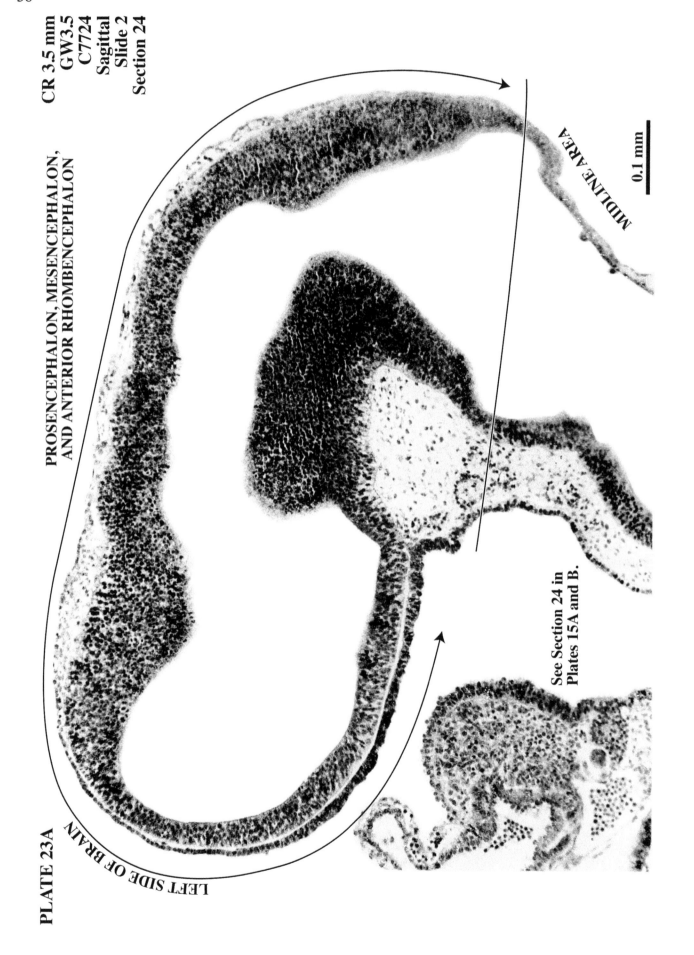

CR 3.5 mm
GW3.5
C7724
Sagittal
Slide 2
Section 24

PLATE 23A

PROSENCEPHALON, MESENCEPHALON,
AND ANTERIOR RHOMBENCEPHALON

LEFT SIDE OF BRAIN

MIDLINE AREA

0.1 mm

See Section 24 in
Plates 15A and B.

PLATE 23B

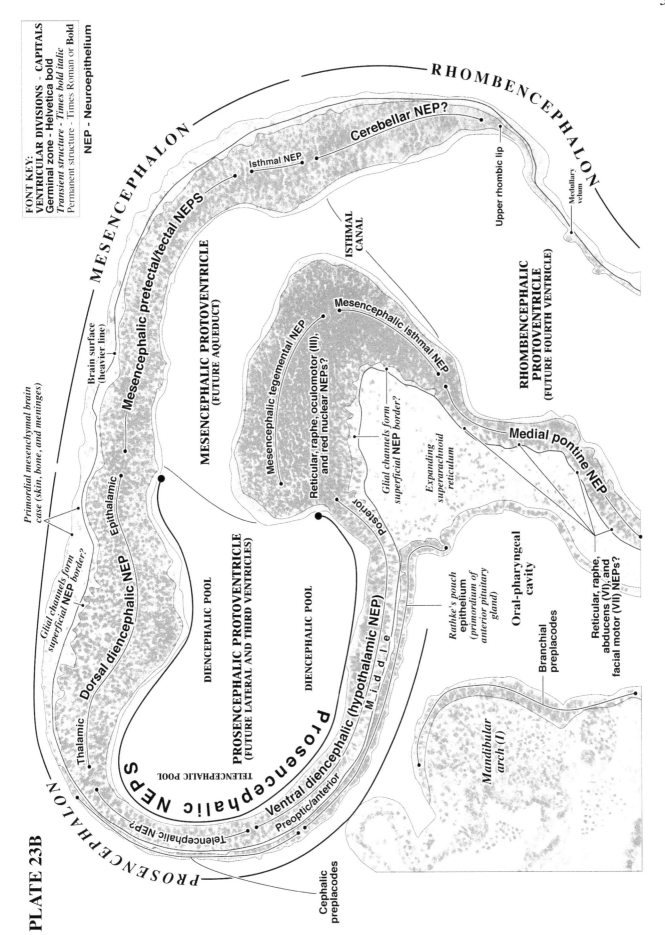

FONT KEY:
VENTRICULAR DIVISIONS - CAPITALS
Germinal zone - **Helvetica bold**
Transient structure - *Times bold italic*
Permanent structure - Times Roman or **Bold**

NEP - Neuroepithelium

CR 3.5 mm. GW3.5
C7724, Sagittal
Slide 2, Section 20

0.1 mm

PLATE 24A

PROSENCEPHALON,
MESENCEPHALON,
AND ANTERIOR
RHOMBENCEPHALON

LEFT SIDE OF BRAIN

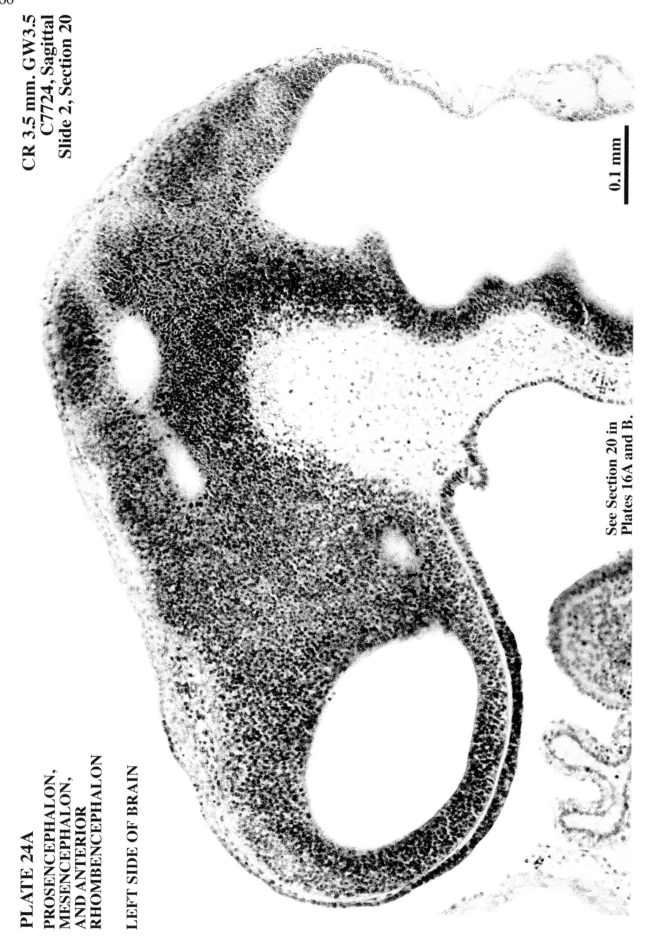

See Section 20 in
Plates 16A and B.

61

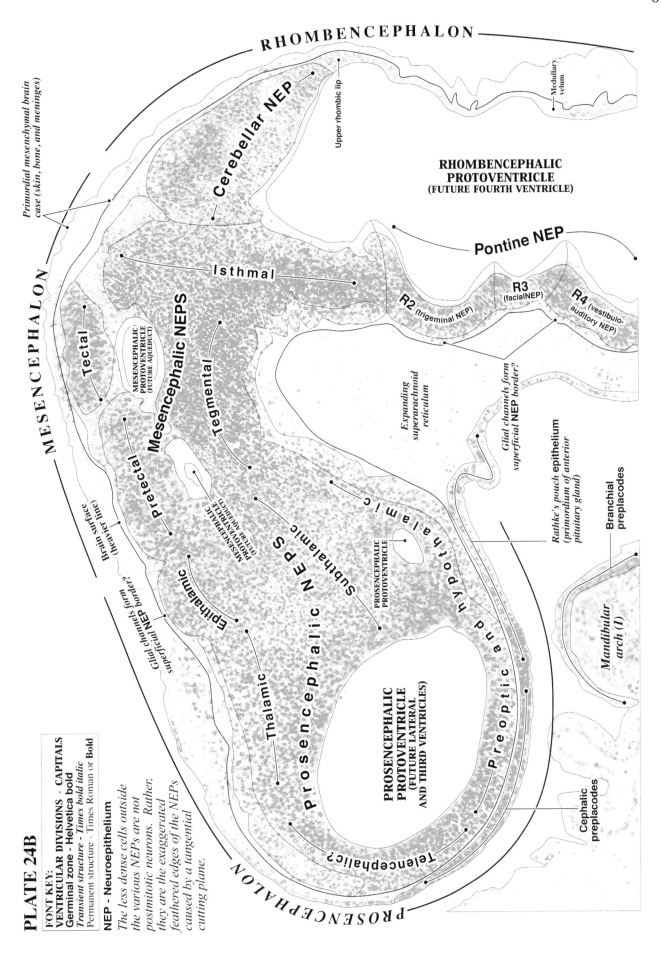

PLATE 24B

FONT KEY:
VENTRICULAR DIVISIONS - CAPITALS
Germinal zone - **Helvetica bold**
Transient structure - *Times bold italic*
Permanent structure - Times Roman or **Bold**

NEP - Neuroepithelium

The less dense cells outside the various NEPs are not postmitotic neurons. Rather, they are the exaggerated feathered edges of the NEPs caused by a tangential cutting plane.

RHOMBENCEPHALON

Cerebellar NEP

Upper rhombic lip

Medullary velum

RHOMBENCEPHALIC PROTOVENTRICLE (FUTURE FOURTH VENTRICLE)

Pontine NEP

Isthmal

R2 (trigeminal NEP)

R3 (facialNEP)

R4 (vestibulo-auditory NEP)

Primordial mesenchymal brain case (skin, bone, and meninges)

MESENCEPHALON

Tectal

MESENCEPHALIC PROTOVENTRICLE (FUTURE AQUEDUCT)

Mesencephalic NEPS

Tegmental

Pretectal

Expanding superarachnoid reticulum

Glial channels form superficial NEP border?

Rathke's pouch epithelium (primordium of anterior pituitary gland)

Branchial preplacodes

Brain surface (heavier line)

Glial channels form superficial NEP border?

Epithalamic

MESENCEPHALIC PROTOVENTRICLE (FUTURE AQUEDUCT)

Prosencephalic NEPS

Thalamic

Subthalamic

Preoptic and hypothalamic

PROSENCEPHALIC PROTOVENTRICLE

PROSENCEPHALIC PROTOVENTRICLE (FUTURE LATERAL AND THIRD VENTRICLES)

Telencephalic?

Mandibular arch (I)

Cephalic preplacodes

PROSENCEPHALON

62

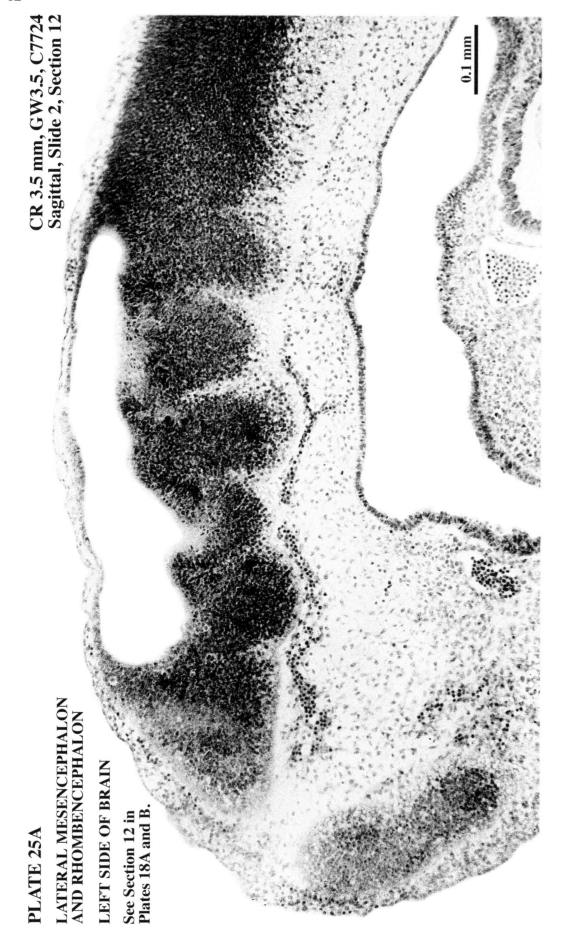

PLATE 25A

LATERAL MESENCEPHALON
AND RHOMBENCEPHALON

LEFT SIDE OF BRAIN

See Section 12 in
Plates 18A and B.

CR 3.5 mm, GW3.5, C7724
Sagittal, Slide 2, Section 12

0.1 mm

63

PLATE 25B

FONT KEY:
VENTRICULAR DIVISIONS - CAPITALS
Germinal zone - Helvetica bold
Transient structure - Times bold italic
Permanent structure - Times Roman or **Bold**

NEP - Neuroepithelium

Arrows indicate the presumed *direction of neuron migration* from neuroepithelial sources.

Primordial mesenchymal brain case (skin, bone, and meninges)

RHOMBENCEPHALON

MESENCEPHALON

Upper rhombic lip
Lower rhombic lip
Cochlear nuclear NEP?
Precerebellar NEP?
Gracile and cuneate nuclear NEP?

Cerebellar NEP?

Isthmal NEP?

Pontine NEP

Medullary NEP

RHOMBENCEPHALIC PROTOVENTRICLE
(FUTURE FOURTH VENTRICLE)

R2 (trigeminal NEP)
R3 (facial NEP)
R4 (vestibulo-auditory NEP)
R5 (vestibulo-auditory NEP)
R6 (glosso-pharyngeal NEP)
R7 (vagal NEP)

Dorsal lower medullary NEP

Outer edge of tectal NEP

Brain surface (heavier line)

Expanding superarachnoid reticulum

The less dense cells outside the various NEPs are not postmitotic neurons. Rather, they are the exaggerated feathered edges of the NEPs caused by a tangential cutting plane.

Oral preplacodes
Cephalic preplacodes
Branchial preplacodes
Oral-pharyngeal cavity
Pharyngeal preplacodes

Mandibular arch (I)
Hyoid arch (II)
Arch III

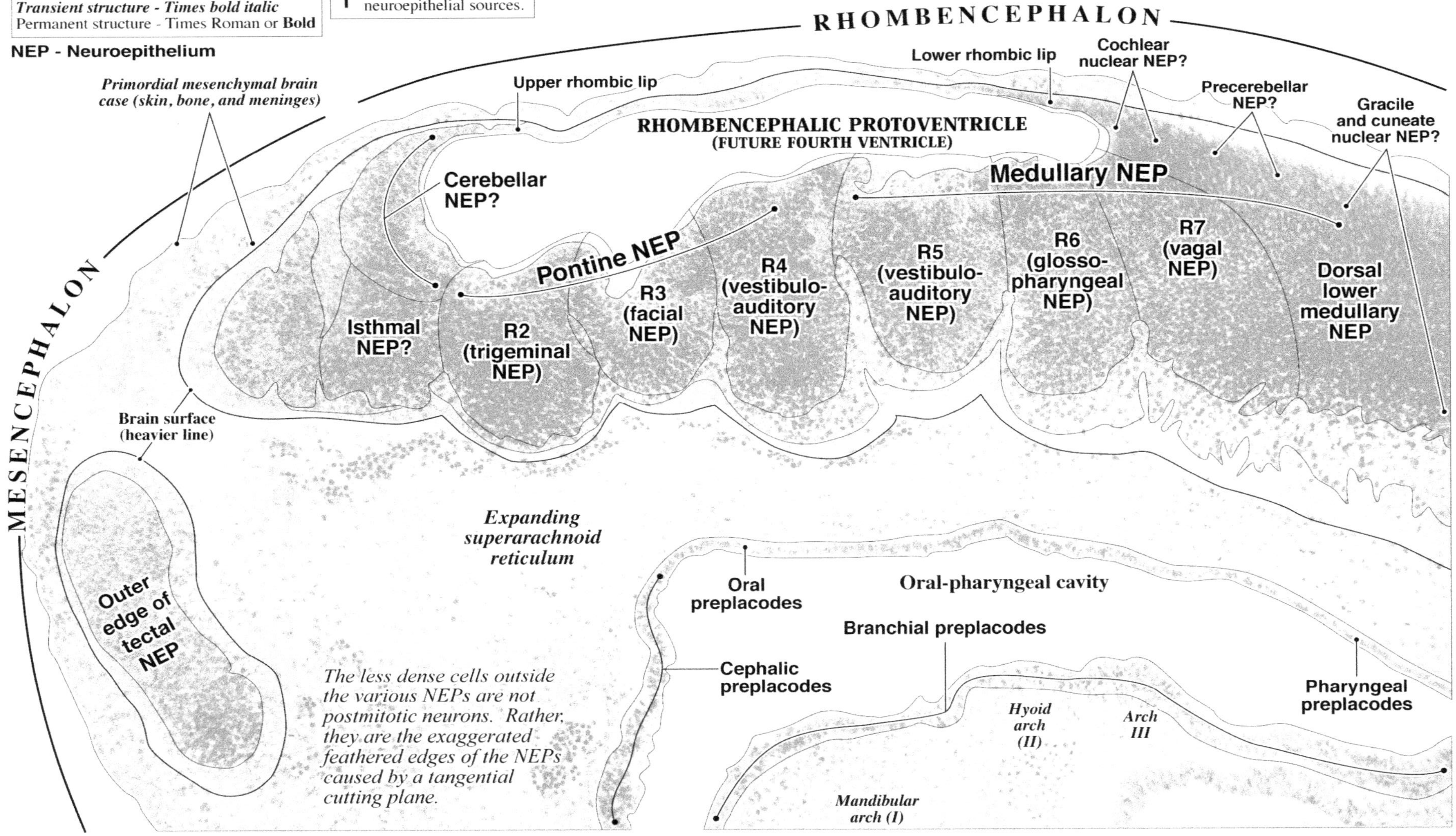

64

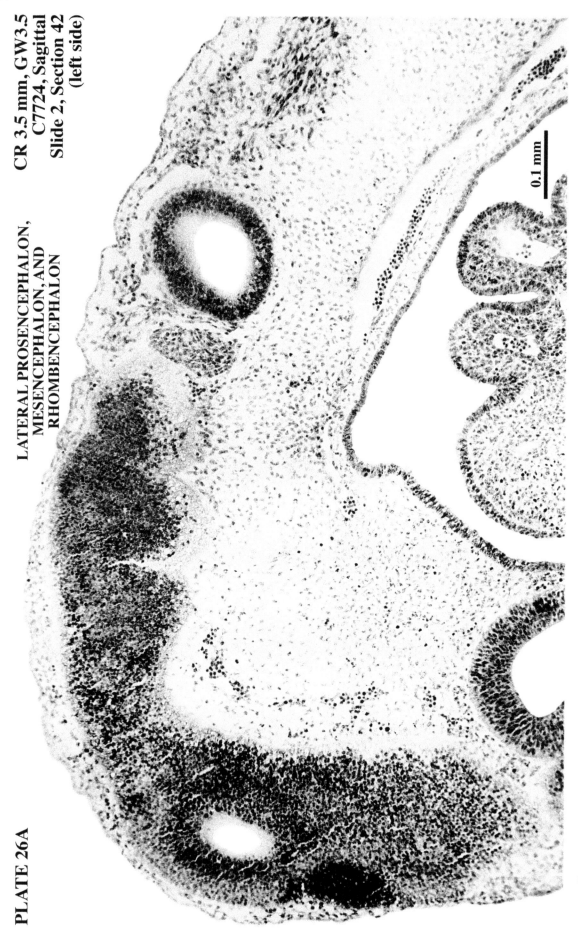

PLATE 26A

LATERAL PROSENCEPHALON,
MESENCEPHALON, AND
RHOMBENCEPHALON

CR 3.5 mm, GW3.5
C7724, Sagittal
Slide 2, Section 42
(left side)

0.1 mm

See similar sections from the right side of the brain
in Plates 19A/B and 20A/B.

PLATE 26B

The less dense cells outside the various NEPs are not postmitotic neurons. Rather, they are the exaggerated feathered edges of the NEPs caused by a tangential cutting plane.

The feathered edge of the various neuroepithelia give the impression that neurons are settling. Actually

NEP - Neuroepithelium

FONT KEY:
VENTRICULAR DIVISIONS - CAPITALS
Germinal zone - Helvetica bold
Transient structure - Times bold italic
Permanent structure - Times Roman or **Bold**

Arrows indicate the presumed *direction of neuron migration* from neuroepithelial sources.

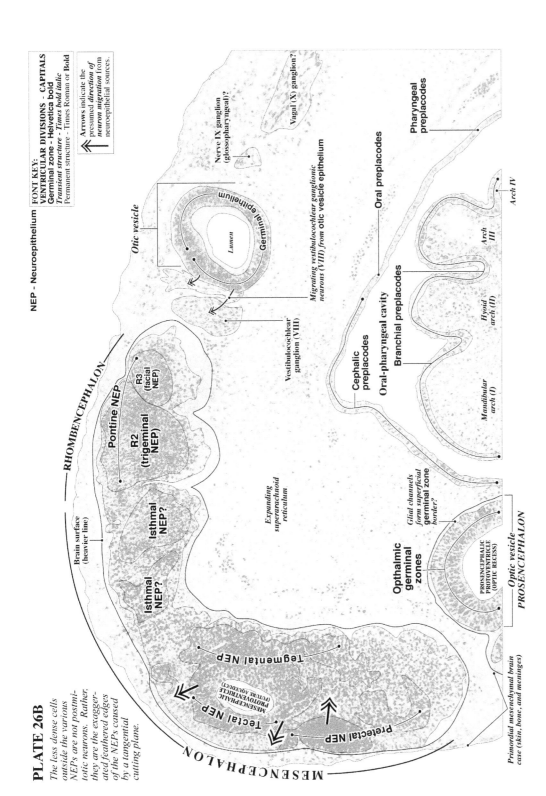

RHOMBENCEPHALON

Brain surface (heavier line)

Otic vesicle

Nerve IX ganglion (glossopharyngeal)?

Vagal (X) ganglion?

Lumen

Germinal epithelium

Pontine NEP

R3 (facial NEP)

R2 (trigeminal NEP)

Migrating vestibulocochlear ganglionic neurons (VIII) from otic vesicle epithellum

Pharyngeal preplacodes

Isthmal NEP?

Vestibulocochlear ganglion (VIII)

Oral preplacodes

Isthmal NEP?

Expanding superarachnoid reticulum

Cephalic preplacodes

Oral-pharyngeal cavity

Branchial preplacodes

Arch III

Arch IV

Tegmental NEP

Glial channels form superficial germinal zone border?

Opthalmic germinal zones

Optic vesicle
PROSENCEPHALON

Hyoid arch (II)

Mandibular arch (I)

Tectal NEP

MESENCEPHALIC PROTOVENTRICLE (FUTURE AQUEDUCT)

Pretectal NEP

PROSENCEPHALIC PROTOVENTRICLE (OPTIC RECESS)

Primordial mexenchymal brain case (skin, bone, and meninges)

MESENCEPHALON

PART IV: C836
CR 4.0 mm (GW 4.0)
Frontal/horizontal

Carnegie Collection specimen #836 (designated here as C836) with a 4-mm crown-rump length (CR) is estimated to be at gestational week (GW) 4. C836 was fixed in corrosive acetic acid, embedded in paraffin, and was cut in 15-μm transverse sections that were stained with aluminum cochineal. Sections of the prosencephalon and anterior mesencephalon are cut in the frontal plane, but the plane shifts to predominantly horizontal in the posterior mesencephalon, pons, and medulla. We photographed 36 sections at low magnification from the frontal prominence to the posterior tips of the mesencephalon and medulla. Twelve of these sections are illustrated in **Plates 27AB to 37AB**. All photographs were used to produce computer-aided 3-D reconstructions of the external features of C836's brain and optic vesicle (**Figure 8**), and to show each illustrated section *in situ* (*insets*, **Plates 27A to 37A**). Each illustrated section shows the brain with all surrounding tissues. Labels in **A Plates** (normal-contrast images) identify non-neural and peripheral neural structures; labels in **B Plates** (low-contrast images) identify central neural structures.

The prosencephalon is still small at GW4.0 and consists of a stockbuilding neuroepithelium surrounding a small prosencephalic superventricle with paired optic recesses. Anterior sections are tentatively identified as future telencephalic neuroepithelium and include the semicircular olfactory placodes at the embryonic surface. But the telencephalic part of the prosencephalon is definitely the smallest part of the brain. The diencephalic neuroepithelium takes up the space just in front of the bilateral optic vesicles. A preplacodal epithelium is in the head around the optic vesicles, but it is continuous with the thickened olfactory placode anterolaterally and the primordium of Rathke's pouch in the ventral midline.

The mesencephalon contains a stockbuilding neuroepithelium surrounding a small mesencephalic superventricle. A roof (tectum) and floor (tegmentum) can be differenti-

ated in coronally cut anterior sections. It is difficult to distinguish neuroepithelial subdivisions in posterior sections that cut the mesencephalon horizontally. The primordial plexiform layer at the brain surface is very thin throughout the entire mesencephalon.

The most prominent neuroepithelial structures in the rhombencephalon are the rhombomere swellings. This specimen is one of the best to show the "rippled" neuroepithelium in **Plates 33** to **36**. There is strong evidence that the rhombomeres are associated with sensory cranial nerves. The trigeminal ganglion (cranial nerve V) is attached to the brain surface at rhombomere 2. The facial ganglion is tentatively identified adjacent to a placode in the hyoid arch that lies immediately ventral to the vestibulocochlear ganglion and posteroventral to rhombomere 3. The vestibulocochlear ganglion (source of VIII nerve) is attached to the rhombomere 4 brain surface. The otic vesicle touches the rhombomere 5 brain surface, and there is cell exchange with R5 NEP. A glossopharyngeal ganglion is lateral to the brain at rhombomere 6. The short nerve extending from the large vagal ganglion (sensory axons of X) touches the rhombomere 7 brain surface. At this early stage, no neurons are migrating from the rhombomeres because stem cells for various populations are in the stockbuilding stage. The feathered basal edge of the NEP is an artifact of the tangential cutting plane. It is possible that some sensory axons from ganglia may invade the outer edges of the rhombomeric NEPs. The small stockbuilding cerebellar neuroepithelium is only identifiable in the most posterior sections of the rhombencephalon and is difficult to distinguish from the mesencephalic tectum. In this specimen, that small cerebellar neuroepithelium is directly adjacent to rhombomere 2 (**Fig. 8, Plates 33AB** to **35AB**). No doubt this spatial relationship prompted some early anatomists to designate the cerebellum as rhombomere 1, but the most anterior rhombomere is still designated as rhombomere 2.

C836 Computer-aided 3-D Brain Reconstructions

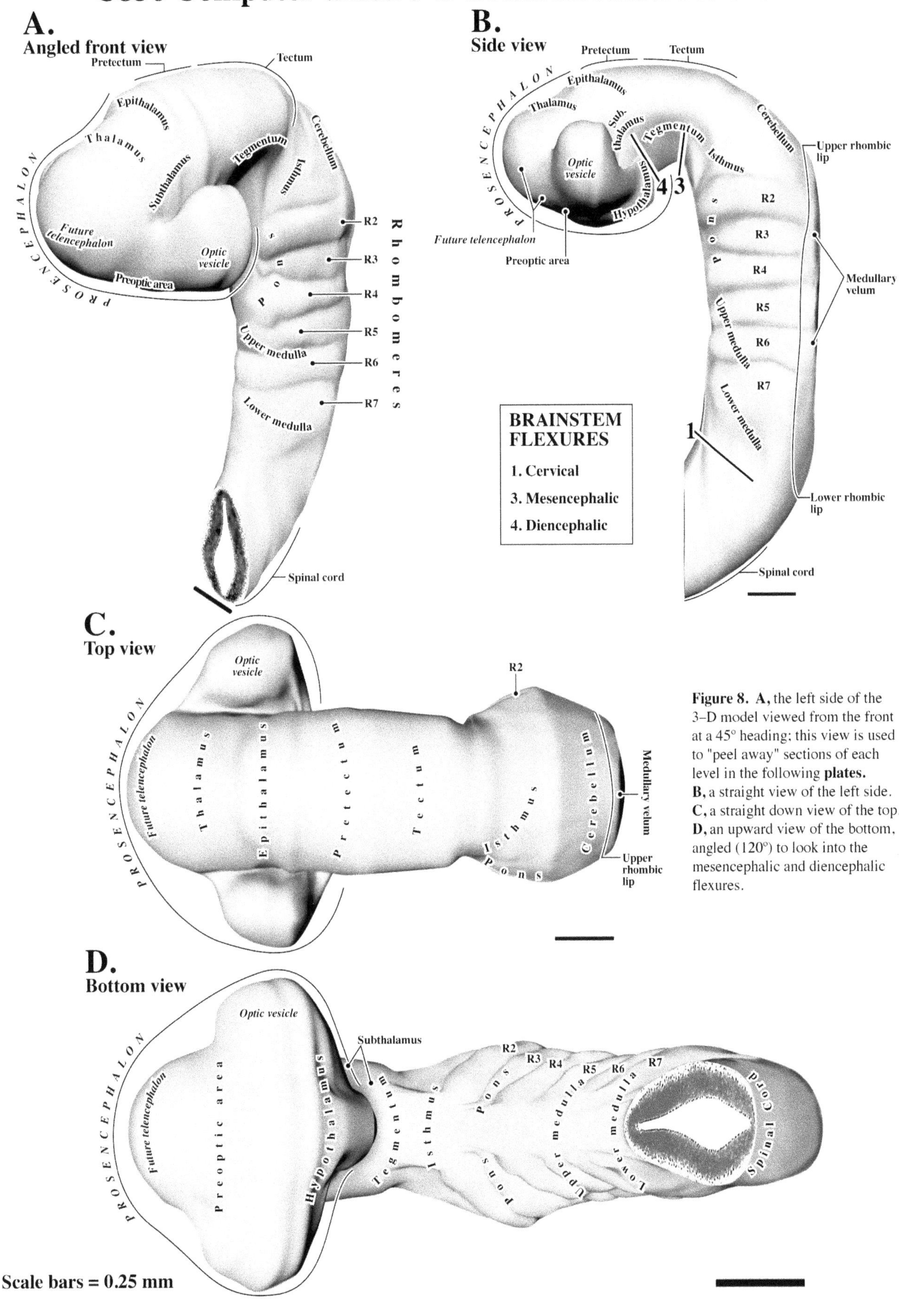

A.
Angled front view

Pretectum
Tectum
Epithalamus
Cerebellum
Thalamus
Tegmentum
Subthalamus
Isthmus
Future telencephalon
Optic vesicle
Preoptic area
P o n s
R2
R3
R4
R5
R6
R7
Upper medulla
Lower medulla
Rhombomeres
Spinal cord

B.
Side view

Pretectum
Tectum
Epithalamus
Thalamus
Sub-thalamus
Tegmentum
Cerebellum
PROSENCEPHALON
Optic vesicle
Isthmus
Upper rhombic lip
4 3
Hypothalamus
R2
R3
R4
R5
R6
R7
Pons
Upper medulla
Medullary velum
Future telencephalon
Preoptic area
Lower medulla
1
Lower rhombic lip
Spinal cord

BRAINSTEM FLEXURES

1. Cervical
3. Mesencephalic
4. Diencephalic

C.
Top view

Optic vesicle
R2
PROSENCEPHALON
Future telencephalon
Thalamus
Epithalamus
Pretectum
Tectum
Isthmus
Pons
Cerebellum
Medullary velum
Upper rhombic lip

Figure 8. A, the left side of the 3–D model viewed from the front at a 45° heading; this view is used to "peel away" sections of each level in the following **plates. B,** a straight view of the left side. **C,** a straight down view of the top. **D,** an upward view of the bottom, angled (120°) to look into the mesencephalic and diencephalic flexures.

D.
Bottom view

Optic vesicle
Subthalamus
PROSENCEPHALON
Future telencephalon
Preoptic area
Hypothalamus
Tegmentum
Isthmus
Pons
R2
R3
R4
R5
R6
R7
Upper medulla
Lower medulla
Spinal cord

Scale bars = 0.25 mm

68

PLATE 27A

CR 4.0 mm
GW4, C836
Frontal/
Horizontal

Peripheral neural and
non-neural structures labeled

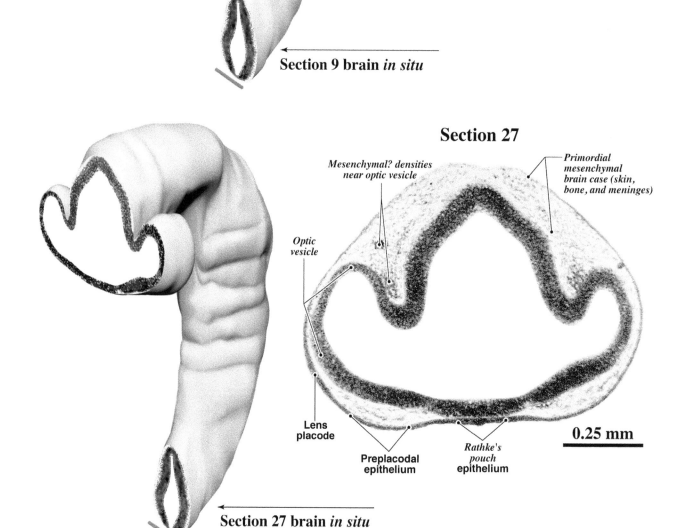

Section 9

Preplacodal
epithelium

Primordial
mesenchymal
brain case (skin,
bone, and meninges)

Olfactory
placode

0.25 mm

Section 9 brain *in situ*

Section 27

Mesenchymal? densities
near optic vesicle

Primordial
mesenchymal
brain case (skin,
bone, and meninges)

Optic
vesicle

Lens
placode

Preplacodal
epithelium

Rathke's
pouch
epithelium

0.25 mm

Section 27 brain *in situ*

Central neural structures labeled

Section 9

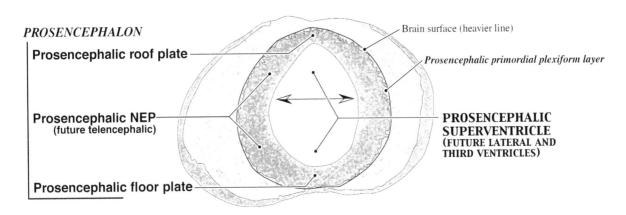

PROSENCEPHALON

Prosencephalic roof plate

Prosencephalic NEP
(future telencephalic)

Prosencephalic floor plate

Brain surface (heavier line)

Prosencephalic primordial plexiform layer

PROSENCEPHALIC SUPERVENTRICLE
(FUTURE LATERAL AND THIRD VENTRICLES)

Section 27

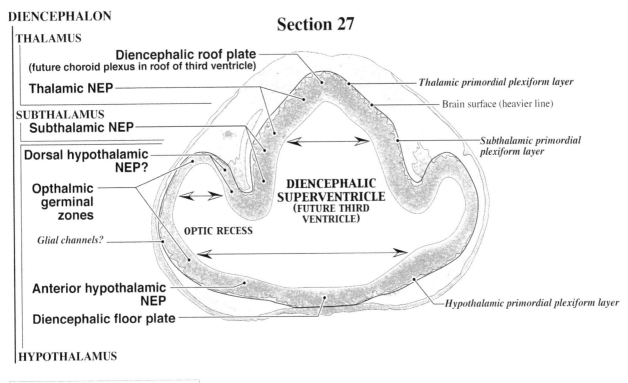

DIENCEPHALON

THALAMUS

Diencephalic roof plate
(future choroid plexus in roof of third ventricle)

Thalamic NEP

SUBTHALAMUS

Subthalamic NEP

Dorsal hypothalamic NEP?

Opthalmic germinal zones

Glial channels?

Anterior hypothalamic NEP

Diencephalic floor plate

HYPOTHALAMUS

Thalamic primordial plexiform layer

Brain surface (heavier line)

Subthalamic primordial plexiform layer

DIENCEPHALIC SUPERVENTRICLE
(FUTURE THIRD VENTRICLE)

OPTIC RECESS

Hypothalamic primordial plexiform layer

FONT KEY:
VENTRICULAR DIVISIONS – CAPITALS
Germinal zone - Helvetica bold
Transient structure - Times bold italic
Permanent structure - Times Roman or **Bold**

ABBREVIATIONS:
GEP - Glioepithelium
NEP - Neuroepithelium

Arrows indicate the regionally *expanding shoreline* of the superventricle with increase in stockbuilding NEP cells.

70

PLATE 28A

Peripheral neural and non-neural structures labeled

CR 4.0 mm, GW4
C836, Frontal/Horizontal
Section 36

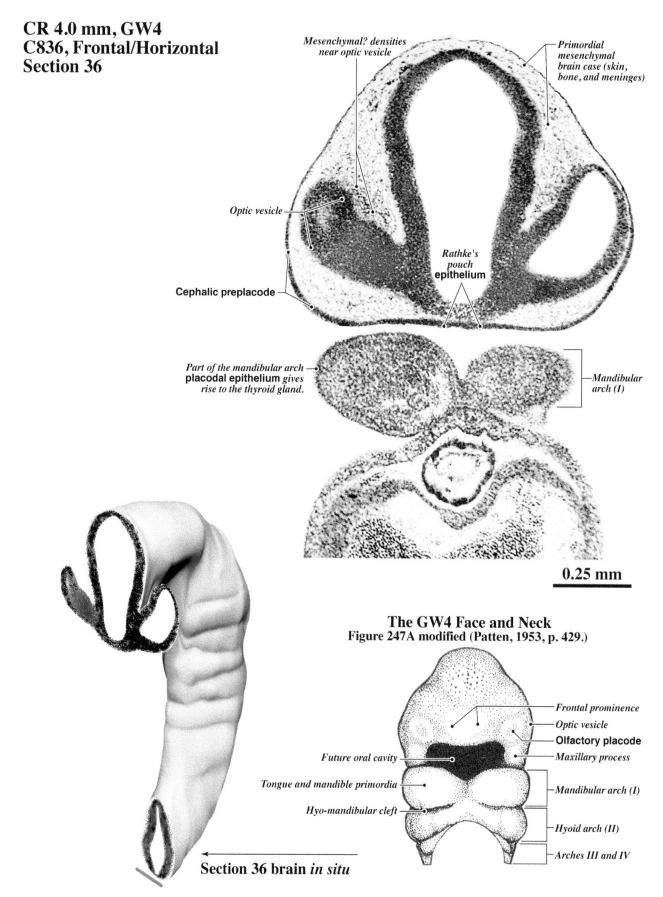

Mesenchymal? densities near optic vesicle

Primordial mesenchymal brain case (skin, bone, and meninges)

Optic vesicle

Rathke's pouch **epithelium**

Cephalic preplacode

Part of the mandibular arch **placodal epithelium** *gives rise to the thyroid gland.*

Mandibular arch (I)

0.25 mm

Section 36 brain *in situ*

The GW4 Face and Neck
Figure 247A modified (Patten, 1953, p. 429.)

Frontal prominence
Optic vesicle
Olfactory placode
Maxillary process
Future oral cavity
Tongue and mandible primordia
Hyo-mandibular cleft
Mandibular arch (I)
Hyoid arch (II)
Arches III and IV

Central neural structures labeled

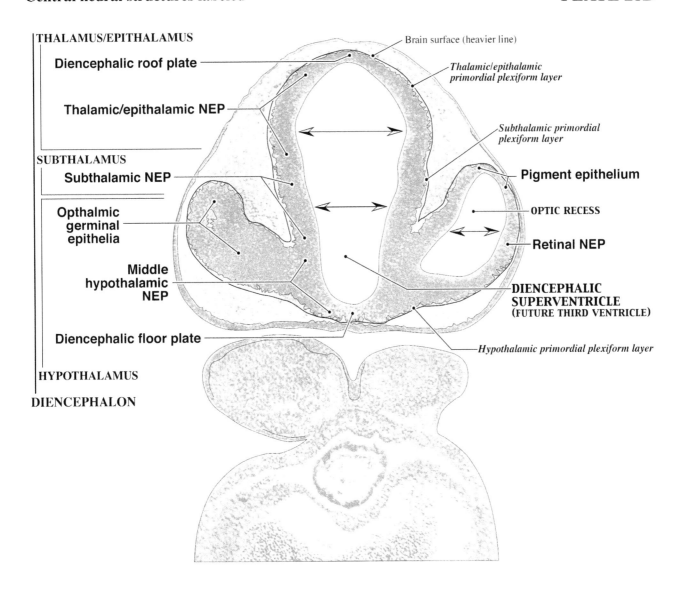

THALAMUS/EPITHALAMUS

Diencephalic roof plate

Thalamic/epithalamic NEP

SUBTHALAMUS

Subthalamic NEP

Opthalmic germinal epithelia

Middle hypothalamic NEP

Diencephalic floor plate

HYPOTHALAMUS

DIENCEPHALON

Brain surface (heavier line)

Thalamic/epithalamic primordial plexiform layer

Subthalamic primordial plexiform layer

Pigment epithelium

OPTIC RECESS

Retinal NEP

DIENCEPHALIC SUPERVENTRICLE
(FUTURE THIRD VENTRICLE)

Hypothalamic primordial plexiform layer

FONT KEY:
VENTRICULAR DIVISIONS – CAPITALS
Germinal zone - Helvetica bold
Transient structure - Times bold italic
Permanent structure - Times Roman or **Bold**

ABBREVIATIONS:
GEP - Glioepithelium
NEP - Neuroepithelium

Arrows indicate the regionally *expanding shoreline* of the superventricle with increase in stockbuilding NEP cells.

PLATE 29A

Peripheral neural and non-neural structures labeled

CR 4.0 mm, GW4
C836, Frontal/Horizontal
Section 48

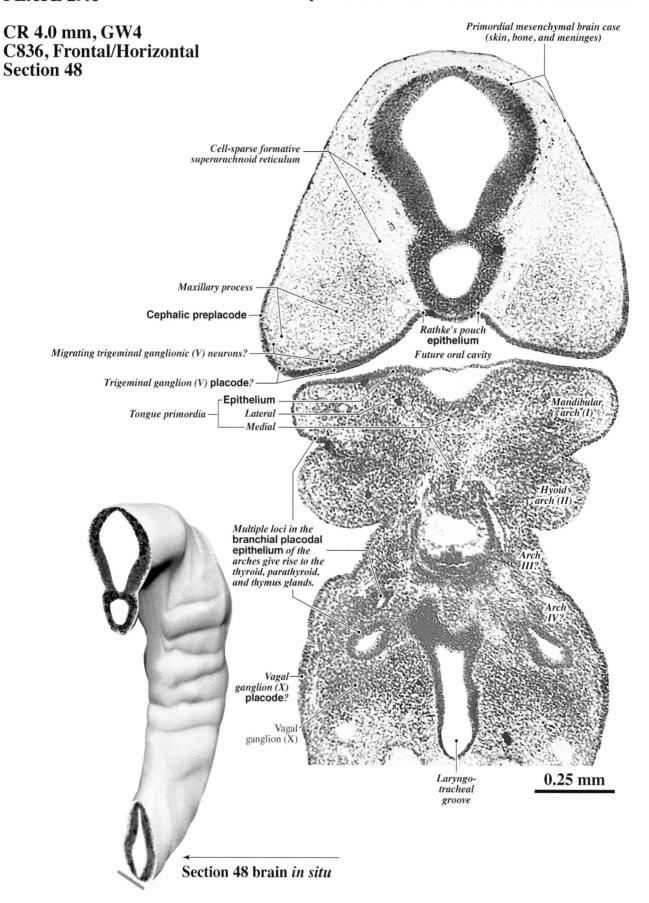

Primordial mesenchymal brain case (skin, bone, and meninges)

Cell-sparse formative superarachnoid reticulum

Maxillary process

Cephalic preplacode

Rathke's pouch epithelium

Future oral cavity

Migrating trigeminal ganglionic (V) neurons?

Trigeminal ganglion (V) placode?

Epithelium

Tongue primordia   *Lateral*

Medial

Mandibular arch (I)

Hyoid arch (II)

Multiple loci in the branchial placodal epithelium *of the arches give rise to the thyroid, parathyroid, and thymus glands.*

Arch III?

Arch IV?

Vagal ganglion (X) placode?

Vagal ganglion (X)

Laryngo-tracheal groove

0.25 mm

Section 48 brain *in situ*

Central neural structures labeled

PLATE 29B

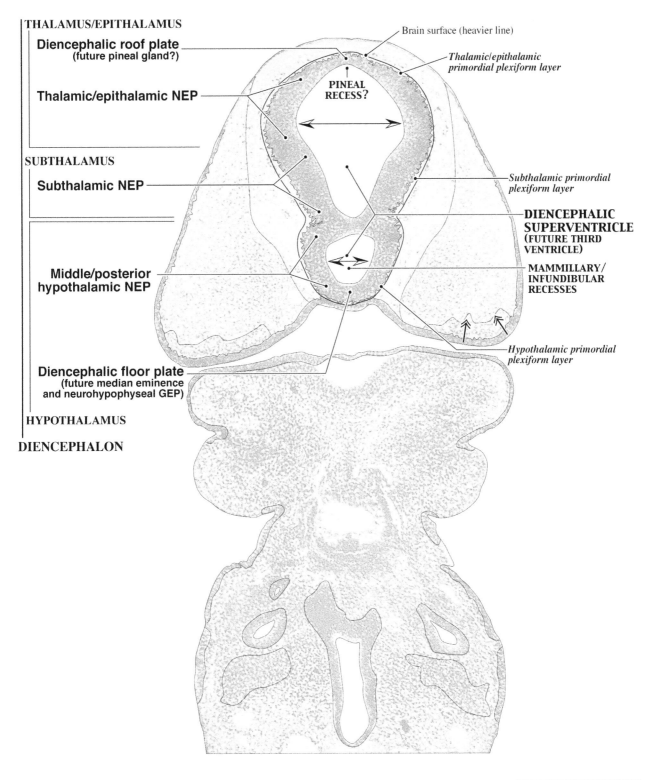

THALAMUS/EPITHALAMUS

Diencephalic roof plate
(future pineal gland?)

Thalamic/epithalamic NEP

SUBTHALAMUS

Subthalamic NEP

**Middle/posterior
hypothalamic NEP**

Diencephalic floor plate
(future median eminence
and neurohypophyseal GEP)

HYPOTHALAMUS

DIENCEPHALON

Brain surface (heavier line)

*Thalamic/epithalamic
primordial plexiform layer*

PINEAL
RECESS?

*Subthalamic primordial
plexiform layer*

**DIENCEPHALIC
SUPERVENTRICLE
(FUTURE THIRD
VENTRICLE)**

**MAMMILLARY/
INFUNDIBULAR
RECESSES**

*Hypothalamic primordial
plexiform layer*

Arrows indicate the
presumed *direction of
neuron migration* from
germinal sources.

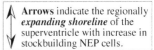

Arrows indicate the regionally
expanding shoreline of the
superventricle with increase in
stockbuilding NEP cells.

FONT KEY:
VENTRICULAR DIVISIONS – CAPITALS
Germinal zone - Helvetica bold
Transient structure - Times bold italic
Permanent structure - Times Roman or **Bold**

ABBREVIATIONS:
GEP - Glioepithelium
NEP - Neuroepithelium

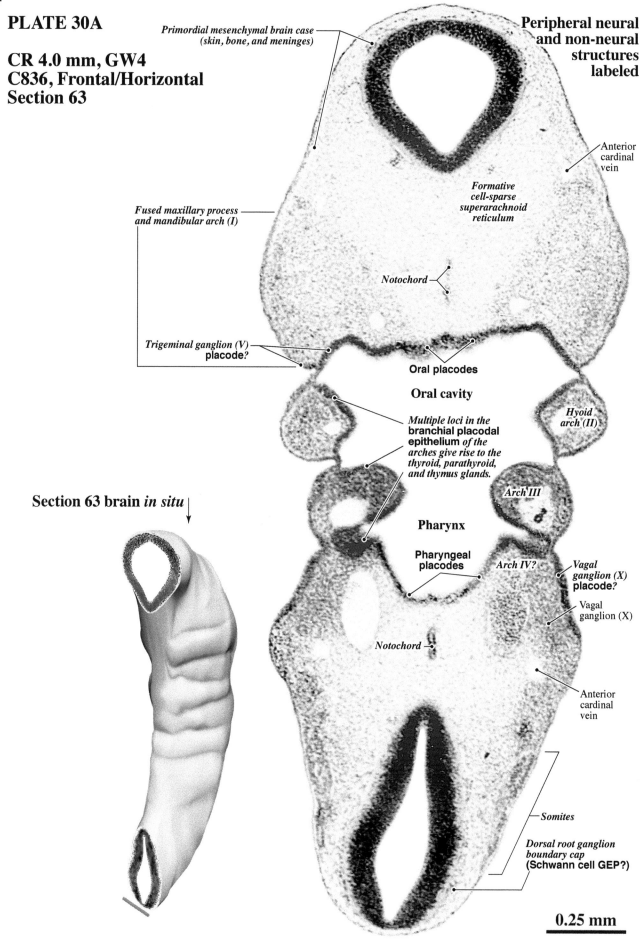

PLATE 30A

**CR 4.0 mm, GW4
C836, Frontal/Horizontal
Section 63**

Peripheral neural
and non-neural
structures
labeled

*Primordial mesenchymal brain case
(skin, bone, and meninges)*

Anterior
cardinal
vein

*Formative
cell-sparse
superarachnoid
reticulum*

*Fused maxillary process
and mandibular arch (I)*

Notochord

Trigeminal ganglion (V)
placode?

Oral placodes

Oral cavity

*Hyoid
arch (II)*

Multiple loci in the
**branchial placodal
epithelium** *of the
arches give rise to the
thyroid, parathyroid,
and thymus glands.*

Arch III

Section 63 brain *in situ*

Pharynx

Pharyngeal
placodes

Arch IV?

*Vagal
ganglion (X)*
placode?

Vagal
ganglion (X)

Notochord

Anterior
cardinal
vein

Somites

*Dorsal root ganglion
boundary cap*
(Schwann cell GEP?)

0.25 mm

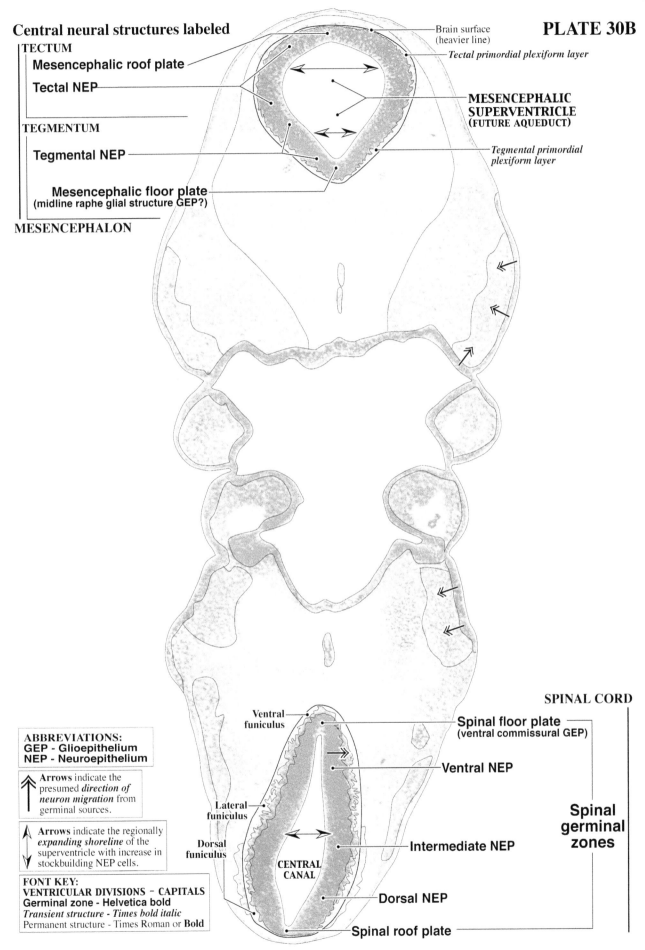

Central neural structures labeled

TECTUM
  Mesencephalic roof plate
  Tectal NEP

TEGMENTUM
  Tegmental NEP

  Mesencephalic floor plate
  (midline raphe glial structure GEP?)

MESENCEPHALON

PLATE 30B

Brain surface
(heavier line)

Tectal primordial plexiform layer

**MESENCEPHALIC
SUPERVENTRICLE
(FUTURE AQUEDUCT)**

*Tegmental primordial
plexiform layer*

SPINAL CORD

Ventral
funiculus

Spinal floor plate
(ventral commissural GEP)

Ventral NEP

Spinal
germinal
zones

Lateral
funiculus

Dorsal
funiculus

**CENTRAL
CANAL**

Intermediate NEP

Dorsal NEP

Spinal roof plate

ABBREVIATIONS:
GEP - Glioepithelium
NEP - Neuroepithelium

Arrows indicate the
presumed *direction of
neuron migration* from
germinal sources.

Arrows indicate the regionally
expanding shoreline of the
superventricle with increase in
stockbuilding NEP cells.

FONT KEY:
VENTRICULAR DIVISIONS – CAPITALS
Germinal zone - Helvetica bold
Transient structure - Times bold italic
Permanent structure - Times Roman or **Bold**

76

PLATE 31A

**CR 4.0 mm, GW4
C836, Frontal/Horizontal
Section 72**

**Peripheral neural
and non-neural
structures
labeled**

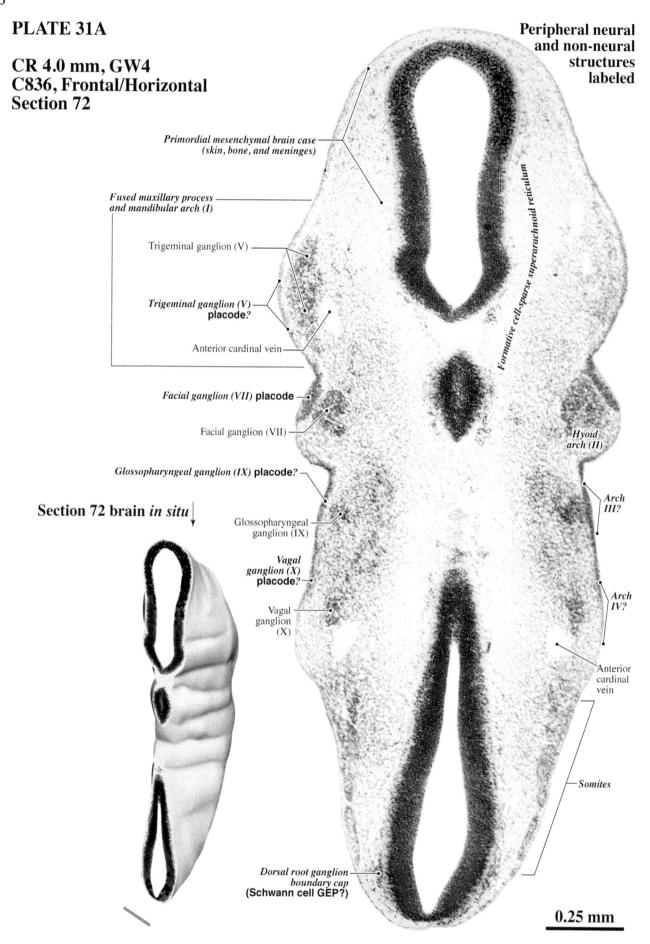

*Primordial mesenchymal brain case
(skin, bone, and meninges)*

*Fused maxillary process
and mandibular arch (I)*

Trigeminal ganglion (V)

Trigeminal ganglion (V)
placode?

Anterior cardinal vein

Facial ganglion (VII) placode

Facial ganglion (VII)

Glossopharyngeal ganglion (IX) placode?

Section 72 brain *in situ* ↓

Glossopharyngeal
ganglion (IX)

*Vagal
ganglion (X)*
placode?

Vagal
ganglion
(X)

*Dorsal root ganglion
boundary cap*
(Schwann cell GEP?)

Formative cell-sparse superarachnoid reticulum

*Hyoid
arch (II)*

*Arch
III?*

*Arch
IV?*

Anterior
cardinal
vein

Somites

0.25 mm

Central neural structures labeled

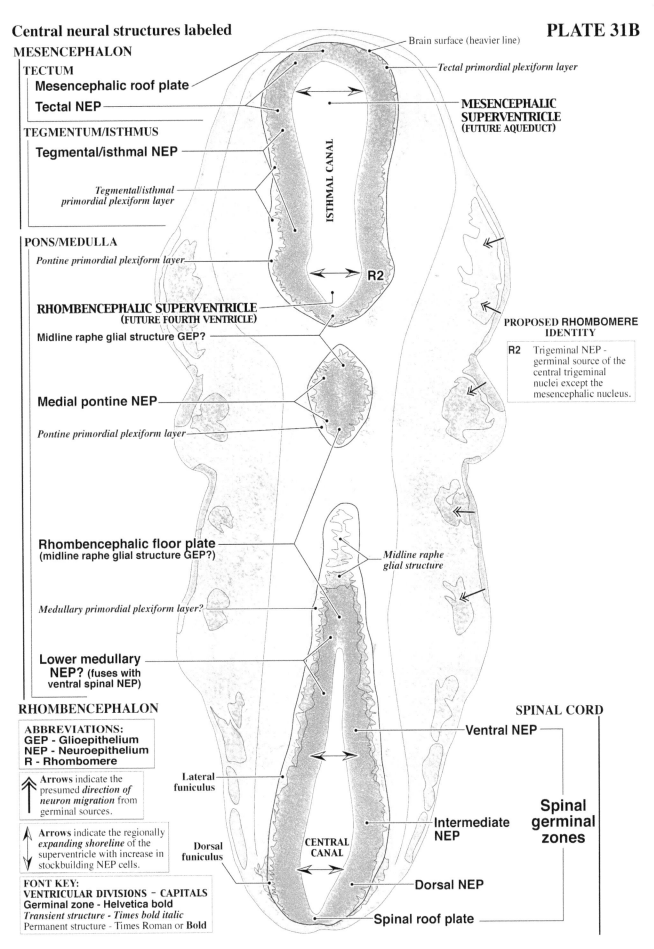

MESENCEPHALON

TECTUM
- **Mesencephalic roof plate**
- **Tectal NEP**

TEGMENTUM/ISTHMUS
- **Tegmental/isthmal NEP**

Tegmental/isthmal primordial plexiform layer

PONS/MEDULLA

Pontine primordial plexiform layer

RHOMBENCEPHALIC SUPERVENTRICLE
(FUTURE FOURTH VENTRICLE)

Midline raphe glial structure GEP?

Medial pontine NEP

Pontine primordial plexiform layer

Rhombencephalic floor plate
(midline raphe glial structure GEP?)

Medullary primordial plexiform layer?

Lower medullary NEP? (fuses with ventral spinal NEP)

RHOMBENCEPHALON

Brain surface (heavier line)

Tectal primordial plexiform layer

MESENCEPHALIC SUPERVENTRICLE
(FUTURE AQUEDUCT)

ISTHMAL CANAL

R2

PROPOSED RHOMBOMERE
IDENTITY

R2   Trigeminal NEP -
germinal source of the
central trigeminal
nuclei except the
mesencephalic nucleus.

Midline raphe glial structure

SPINAL CORD

Ventral NEP

Spinal germinal zones

Lateral funiculus

Dorsal funiculus

CENTRAL CANAL

Intermediate NEP

Dorsal NEP

Spinal roof plate

ABBREVIATIONS:
GEP - Glioepithelium
NEP - Neuroepithelium
R - Rhombomere

Arrows indicate the
presumed *direction of
neuron migration* from
germinal sources.

Arrows indicate the regionally
expanding shoreline of the
superventricle with increase in
stockbuilding NEP cells.

FONT KEY:
VENTRICULAR DIVISIONS - CAPITALS
Germinal zone - Helvetica bold
Transient structure - Times bold italic
Permanent structure - Times Roman or **Bold**

PLATE 32A

CR 4.0 mm, GW4
C836, Frontal/Horizontal
Section 75

Primordial mesenchymal brain case
(skin, bone, and meninges)

Peripheral neural
and non-neural
structures
labeled

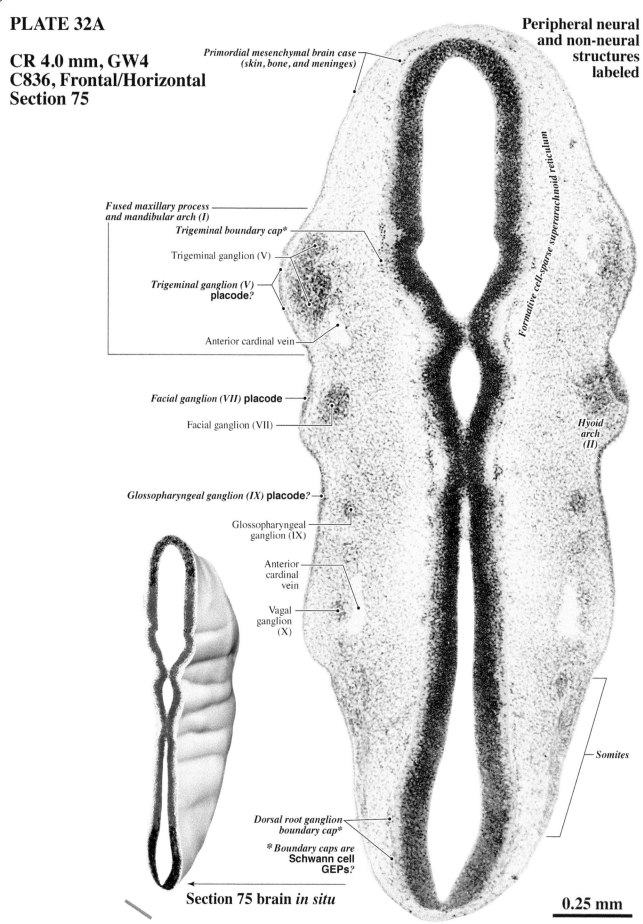

Formative cell-sparse superarachnoid reticulum

Fused maxillary process
and mandibular arch (I)

Trigeminal boundary cap*

Trigeminal ganglion (V)

Trigeminal ganglion (V)
placode?

Anterior cardinal vein

Facial ganglion (VII) placode

Facial ganglion (VII)

Hyoid
arch
(II)

Glossopharyngeal ganglion (IX) placode?

Glossopharyngeal
ganglion (IX)

Anterior
cardinal
vein

Vagal
ganglion
(X)

Somites

Dorsal root ganglion
boundary cap*

* Boundary caps are
Schwann cell
GEPs?

← **Section 75 brain in situ**

0.25 mm

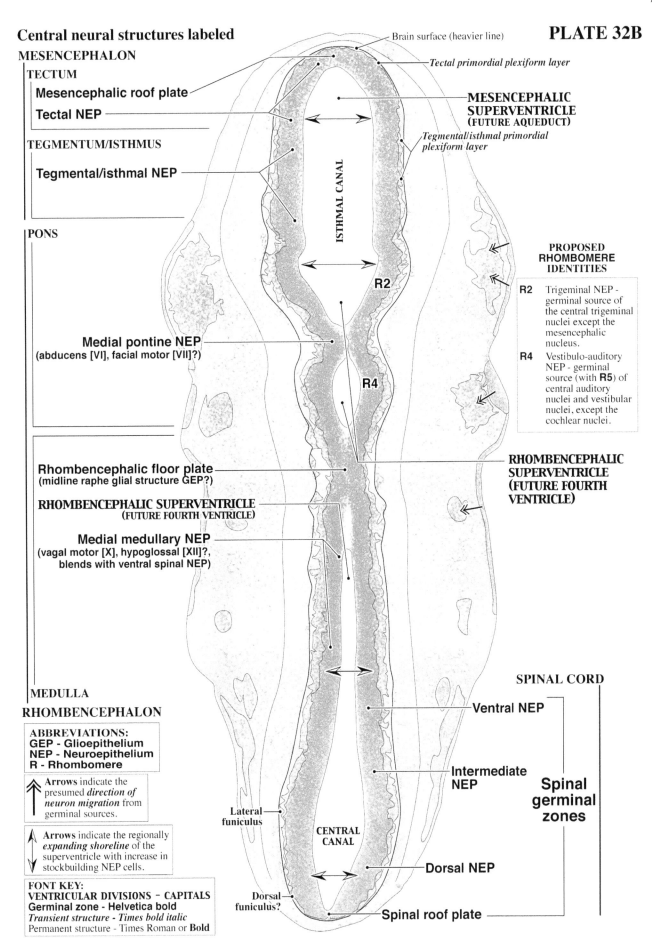

Central neural structures labeled   **PLATE 32B**

Brain surface (heavier line)

MESENCEPHALON
  TECTUM
    Mesencephalic roof plate
    Tectal NEP
  TEGMENTUM/ISTHMUS
    Tegmental/isthmal NEP

PONS

    Medial pontine NEP
    (abducens [VI], facial motor [VII]?)

    Rhombencephalic floor plate
    (midline raphe glial structure GEP?)

    RHOMBENCEPHALIC SUPERVENTRICLE
    (FUTURE FOURTH VENTRICLE)

    Medial medullary NEP
    (vagal motor [X], hypoglossal [XII]?,
    blends with ventral spinal NEP)

MEDULLA
RHOMBENCEPHALON

Tectal primordial plexiform layer

**MESENCEPHALIC
SUPERVENTRICLE
(FUTURE AQUEDUCT)**

*Tegmental/isthmal primordial
plexiform layer*

ISTHMAL CANAL

R2

R4

**PROPOSED
RHOMBOMERE
IDENTITIES**

R2 Trigeminal NEP - germinal source of the central trigeminal nuclei except the mesencephalic nucleus.

R4 Vestibulo-auditory NEP - germinal source (with R5) of central auditory nuclei and vestibular nuclei, except the cochlear nuclei.

**RHOMBENCEPHALIC
SUPERVENTRICLE
(FUTURE FOURTH
VENTRICLE)**

SPINAL CORD

Ventral NEP

Intermediate NEP

**Spinal
germinal
zones**

Lateral funiculus

CENTRAL CANAL

Dorsal NEP

Dorsal funiculus?

Spinal roof plate

ABBREVIATIONS:
GEP - Glioepithelium
NEP - Neuroepithelium
R - Rhombomere

Arrows indicate the presumed *direction of neuron migration* from germinal sources.

Arrows indicate the regionally *expanding shoreline* of the superventricle with increase in stockbuilding NEP cells.

FONT KEY:
VENTRICULAR DIVISIONS - CAPITALS
Germinal zone - Helvetica bold
Transient structure - Times bold italic
Permanent structure - Times Roman or **Bold**

80

PLATE 33A

CR 4.0 mm, GW4
C836, Frontal/Horizontal
Section 81

Peripheral neural and non-neural
structures labeled

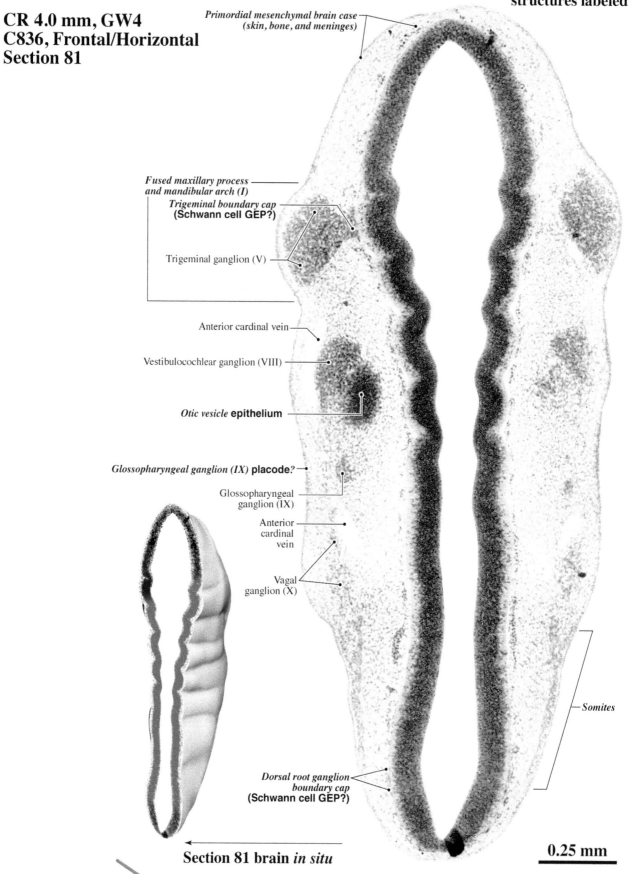

Primordial mesenchymal brain case
(skin, bone, and meninges)

Fused maxillary process
and mandibular arch (I)

Trigeminal boundary cap
(Schwann cell GEP?)

Trigeminal ganglion (V)

Anterior cardinal vein

Vestibulocochlear ganglion (VIII)

Otic vesicle **epithelium**

Glossopharyngeal ganglion (IX) **placode?**

Glossopharyngeal
ganglion (IX)

Anterior
cardinal
vein

Vagal
ganglion (X)

Somites

Dorsal root ganglion
boundary cap
(Schwann cell GEP?)

← **Section 81 brain** *in situ*

0.25 mm

Central neural structures labeled

MESENCEPHALON

TECTUM?

Mesencephalic roof plate

Posterior tip of tectal NEP?

ISTHMUS

Isthmal NEP

CEREBELLUM

Cerebellar NEP?

Fibrous layer in superficial cerebellum?

PONS

Brain surface (heavier line)

Tectal primordial plexiform layer?

ISTHMAL CANAL

MESENCEPHALIC SUPERVENTRICLE (FUTURE AQUEDUCT)

Isthmal primordial plexiform layer

R2

R3

R4

R5

R6

R7

RHOMBENCEPHALIC SUPERVENTRICLE (FUTURE FOURTH VENTRICLE)

Lower intermediate medullary NEP (blends with intermediate spinal NEP)

MEDULLA

RHOMBENCEPHALON

PROPOSED RHOMBOMERE IDENTITIES

R2	Trigeminal NEP - germinal source of the central trigeminal nuclei except the mesencephalic nucleus.
R3	Facial NEP - germinal source of central sensory neurons getting input from the facial (VII) ganglion.
R4	Vestibulo-auditory NEP - germinal source (with **R5**) of central auditory nuclei and vestibular nuclei, except the cochlear nuclei.
R5	Vestibulo-auditory NEP - germinal source (with **R4**) of central auditory nuclei and vestibular nuclei, except the cochlear nuclei.
R6	Glossopharyngeal NEP - germinal source of sensory neurons that receive input from the glossopharyngeal (IX) ganglion.
R7	Vagal (X) sensory NEP - germinal source of the dorsal sensory nucleus and other sensory vagal nuclei.

ABBREVIATIONS:
GEP - Glioepithelium
NEP - Neuroepithelium
R - Rhombomere

Arrows indicate the presumed *direction of neuron migration* from germinal sources.

Arrows indicate the regionally *expanding shoreline* of the superventricle with increase in stockbuilding NEP cells.

FONT KEY:
VENTRICULAR DIVISIONS – CAPITALS
Germinal zone - Helvetica bold
Transient structure - Times bold italic
Permanent structure - Times Roman or **Bold**

Lateral funiculus

Dorsal funiculus?

CENTRAL CANAL

SPINAL CORD

Intermediate NEP

Spinal germinal zones

Dorsal NEP

Spinal roof plate

82

CR 4.0 mm, GW4
C836, Frontal/Horizontal
Section 84

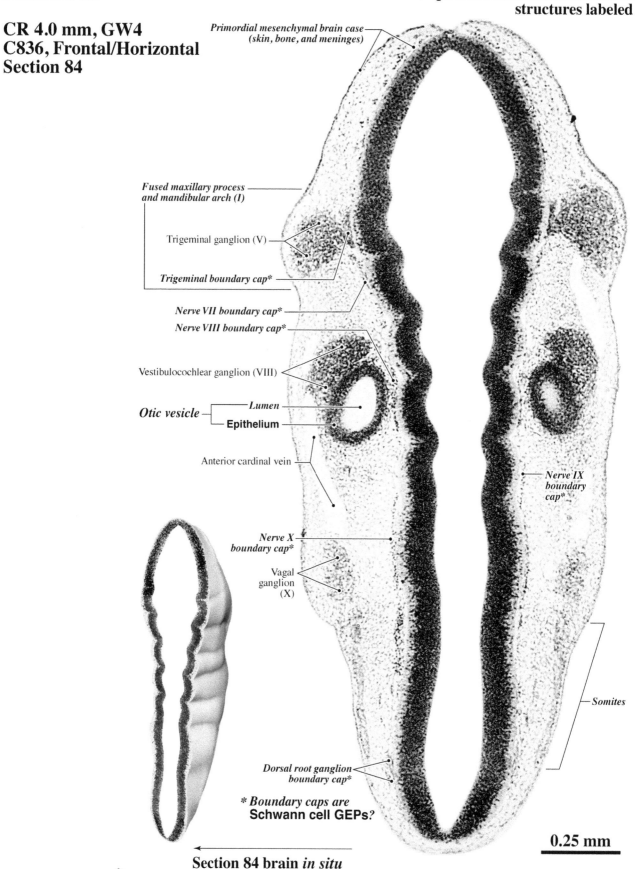

Primordial mesenchymal brain case
(skin, bone, and meninges)

Fused maxillary process
and mandibular arch (I)

Trigeminal ganglion (V)

Trigeminal boundary cap*

Nerve VII boundary cap*

Nerve VIII boundary cap*

Vestibulocochlear ganglion (VIII)

Otic vesicle — Lumen

Epithelium

Anterior cardinal vein

Nerve IX
boundary
cap*

Nerve X
boundary cap*

Vagal
ganglion
(X)

Somites

Dorsal root ganglion
boundary cap*

* Boundary caps are
Schwann cell GEPs?

0.25 mm

Section 84 brain in situ

Central neural structures labeled

Brain surface (heavier line)

Tectal primordial plexiform layer?

MESENCEPHALON

TECTUM?

Mesencephalic roof plate

Posterior tip of tectal NEP?

ISTHMAL CANAL

MESENCEPHALIC SUPERVENTRICLE (FUTURE AQUEDUCT)

ISTHMUS

Isthmal NEP

Isthmal primordial plexiform layer

CEREBELLUM

Cerebellar NEP

Fibrous layer in superficial cerebellum

PONS

R2

R3

PROPOSED RHOMBOMERE IDENTITIES

R2	Trigeminal NEP - germinal source of the central trigeminal nuclei except the mesencephalic nucleus.
R3	Facial NEP - germinal source of central sensory neurons getting input from the facial (VII) ganglion.
R4	Vestibulo-auditory NEP - germinal source (with **R5**) of central auditory nuclei and vestibular nuclei, except the cochlear nuclei.
R5	Vestibulo-auditory NEP - germinal source (with **R4**) of central auditory nuclei and vestibular nuclei, except the cochlear nuclei.
R6	Glossopharyngeal NEP - germinal source of sensory neurons that receive input from the glossopharyngeal (IX) ganglion.
R7	Vagal (X) sensory NEP - germinal source of the dorsal sensory nucleus and other sensory vagal nuclei.

R4

R5

R6

R7

RHOMBENCEPHALIC SUPERVENTRICLE (FUTURE FOURTH VENTRICLE)

Lower intermediate medullary NEP
(blends with intermediate spinal NEP)

MEDULLA

RHOMBENCEPHALON

ABBREVIATIONS:
GEP - Glioepithelium
NEP - Neuroepithelium
R - Rhombomere

Arrows indicate the presumed *direction of neuron migration* from germinal sources.

Arrows indicate the regionally *expanding shoreline* of the superventricle with increase in stockbuilding NEP cells.

FONT KEY:
VENTRICULAR DIVISIONS - CAPITALS
Germinal zone - Helvetica bold
Transient structure - Times bold italic
Permanent structure - Times Roman or **Bold**

Lateral funiculus

Dorsal funiculus?

CENTRAL CANAL

SPINAL CORD

Intermediate NEP

Spinal germinal zones

Dorsal NEP

Spinal roof plate

PLATE 35A

**Peripheral neural and non-neural
structures labeled**

**CR 4.0 mm, GW4
C836, Frontal/Horizontal
Section 90**

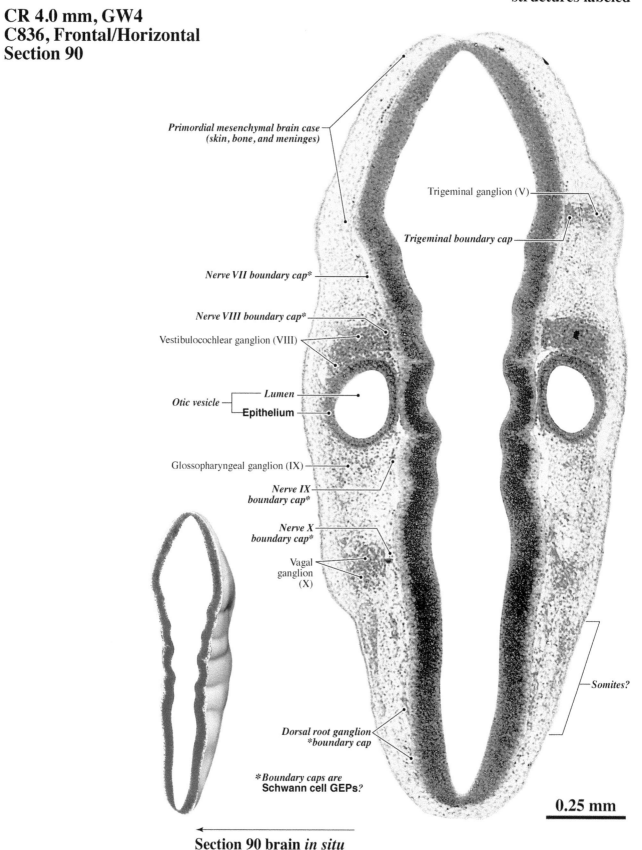

*Primordial mesenchymal brain case
(skin, bone, and meninges)*

Trigeminal ganglion (V)

Trigeminal boundary cap

Nerve VII boundary cap*

Nerve VIII boundary cap*

Vestibulocochlear ganglion (VIII)

Lumen

Otic vesicle ⌐
Epithelium

Glossopharyngeal ganglion (IX)

*Nerve IX
boundary cap**

*Nerve X
boundary cap**

Vagal
ganglion
(X)

Somites?

*Dorsal root ganglion
boundary cap

**Boundary caps are*
Schwann cell GEPs?

Section 90 brain *in situ*

0.25 mm

Central neural structures labeled

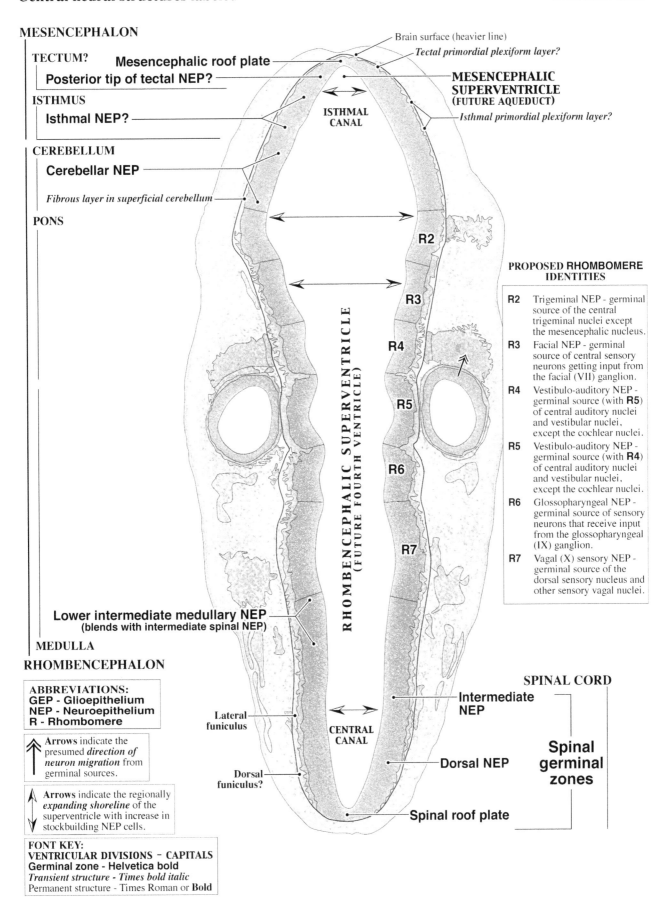

MESENCEPHALON

TECTUM?

Mesencephalic roof plate

Brain surface (heavier line)

Tectal primordial plexiform layer?

Posterior tip of tectal NEP?

MESENCEPHALIC SUPERVENTRICLE
(FUTURE AQUEDUCT)

ISTHMUS

Isthmal NEP?

ISTHMAL CANAL

Isthmal primordial plexiform layer?

CEREBELLUM

Cerebellar NEP

Fibrous layer in superficial cerebellum

PONS

R2

R3

R4

R5

R6

R7

RHOMBENCEPHALIC SUPERVENTRICLE
(FUTURE FOURTH VENTRICLE)

Lower intermediate medullary NEP
(blends with intermediate spinal NEP)

MEDULLA

RHOMBENCEPHALON

PROPOSED RHOMBOMERE IDENTITIES

R2 Trigeminal NEP - germinal source of the central trigeminal nuclei except the mesencephalic nucleus.

R3 Facial NEP - germinal source of central sensory neurons getting input from the facial (VII) ganglion.

R4 Vestibulo-auditory NEP - germinal source (with **R5**) of central auditory nuclei and vestibular nuclei, except the cochlear nuclei.

R5 Vestibulo-auditory NEP - germinal source (with **R4**) of central auditory nuclei and vestibular nuclei, except the cochlear nuclei.

R6 Glossopharyngeal NEP - germinal source of sensory neurons that receive input from the glossopharyngeal (IX) ganglion.

R7 Vagal (X) sensory NEP - germinal source of the dorsal sensory nucleus and other sensory vagal nuclei.

ABBREVIATIONS:
GEP - Glioepithelium
NEP - Neuroepithelium
R - Rhombomere

Arrows indicate the presumed *direction of neuron migration* from germinal sources.

Arrows indicate the regionally *expanding shoreline* of the superventricle with increase in stockbuilding NEP cells.

FONT KEY:
VENTRICULAR DIVISIONS – CAPITALS
Germinal zone - Helvetica bold
Transient structure - Times bold italic
Permanent structure - Times Roman or **Bold**

Lateral funiculus

CENTRAL CANAL

Dorsal funiculus?

SPINAL CORD

Intermediate NEP

Spinal germinal zones

Dorsal NEP

Spinal roof plate

PLATE 36A

**Peripheral neural and non-neural
structures labeled**

CR 4.0 mm, GW4
C836, Frontal/Horizontal
Section 93

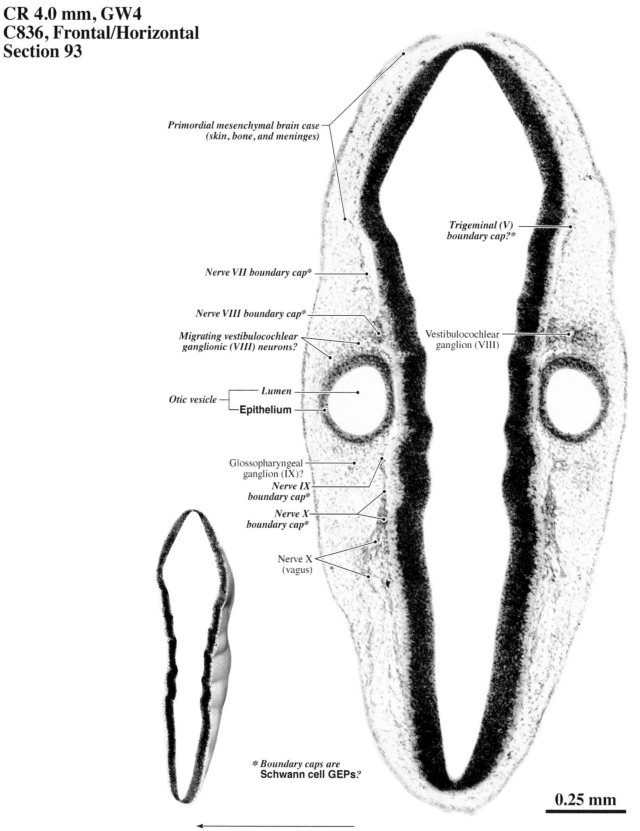

Primordial mesenchymal brain case
(skin, bone, and meninges)

Trigeminal (V)
boundary cap?*

Nerve VII boundary cap*

Nerve VIII boundary cap*

Migrating vestibulocochlear
ganglionic (VIII) neurons?

Vestibulocochlear
ganglion (VIII)

Lumen

Otic vesicle

Epithelium

Glossopharyngeal
ganglion (IX)?

Nerve IX
boundary cap*

Nerve X
boundary cap*

Nerve X
(vagus)

*Boundary caps are
Schwann cell GEPs?

0.25 mm

Section 93 brain *in situ*

Central neural structures labeled

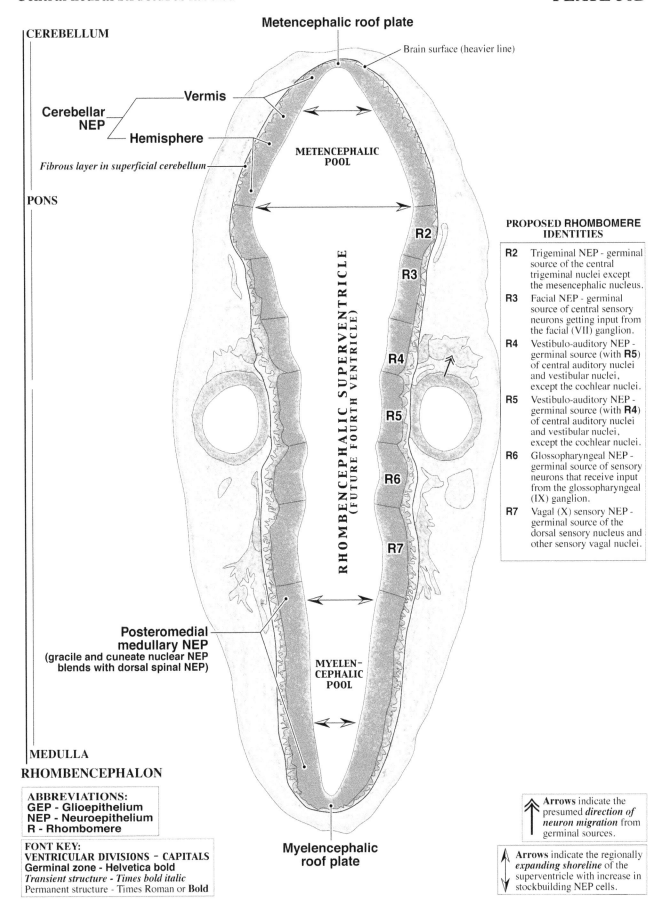

Metencephalic roof plate

Brain surface (heavier line)

CEREBELLUM

Vermis

Cerebellar NEP

Hemisphere

Fibrous layer in superficial cerebellum

METENCEPHALIC POOL

PONS

R2

R3

R4

R5

R6

R7

RHOMBENCEPHALIC SUPERVENTRICLE
(FUTURE FOURTH VENTRICLE)

PROPOSED RHOMBOMERE IDENTITIES

R2	Trigeminal NEP - germinal source of the central trigeminal nuclei except the mesencephalic nucleus.
R3	Facial NEP - germinal source of central sensory neurons getting input from the facial (VII) ganglion.
R4	Vestibulo-auditory NEP - germinal source (with **R5**) of central auditory nuclei and vestibular nuclei, except the cochlear nuclei.
R5	Vestibulo-auditory NEP - germinal source (with **R4**) of central auditory nuclei and vestibular nuclei, except the cochlear nuclei.
R6	Glossopharyngeal NEP - germinal source of sensory neurons that receive input from the glossopharyngeal (IX) ganglion.
R7	Vagal (X) sensory NEP - germinal source of the dorsal sensory nucleus and other sensory vagal nuclei.

Posteromedial medullary NEP
(gracile and cuneate nuclear NEP blends with dorsal spinal NEP)

MYELEN- CEPHALIC POOL

MEDULLA

RHOMBENCEPHALON

ABBREVIATIONS:
GEP - Glioepithelium
NEP - Neuroepithelium
R - Rhombomere

FONT KEY:
VENTRICULAR DIVISIONS – CAPITALS
Germinal zone - Helvetica bold
Transient structure - Times bold italic
Permanent structure - Times Roman or **Bold**

Myelencephalic roof plate

Arrows indicate the presumed *direction of neuron migration* from germinal sources.

Arrows indicate the regionally *expanding shoreline* of the superventricle with increase in stockbuilding NEP cells.

PLATE 37A

CR 4.0 mm, GW4
C836, Frontal/Horizontal
Section 99

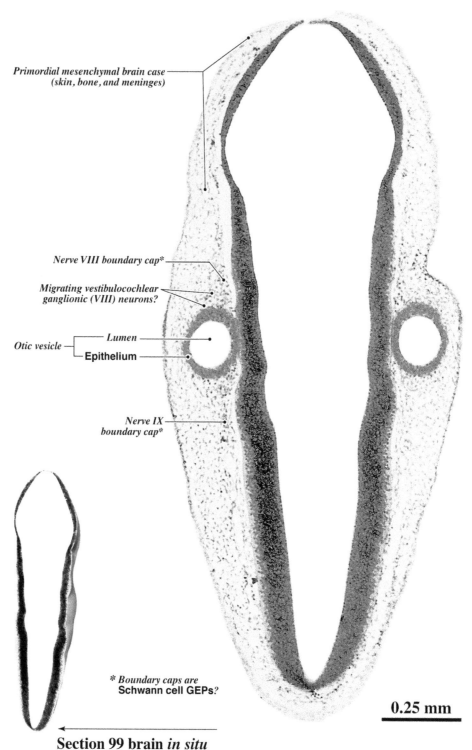

Primordial mesenchymal brain case
(skin, bone, and meninges)

Nerve VIII boundary cap*

Migrating vestibulocochlear
ganglionic (VIII) neurons?

Lumen

Otic vesicle **Epithelium**

Nerve IX
boundary cap*

* *Boundary caps are*
Schwann cell GEPs*?*

0.25 mm

Section 99 brain *in situ*

Central neural structures labeled

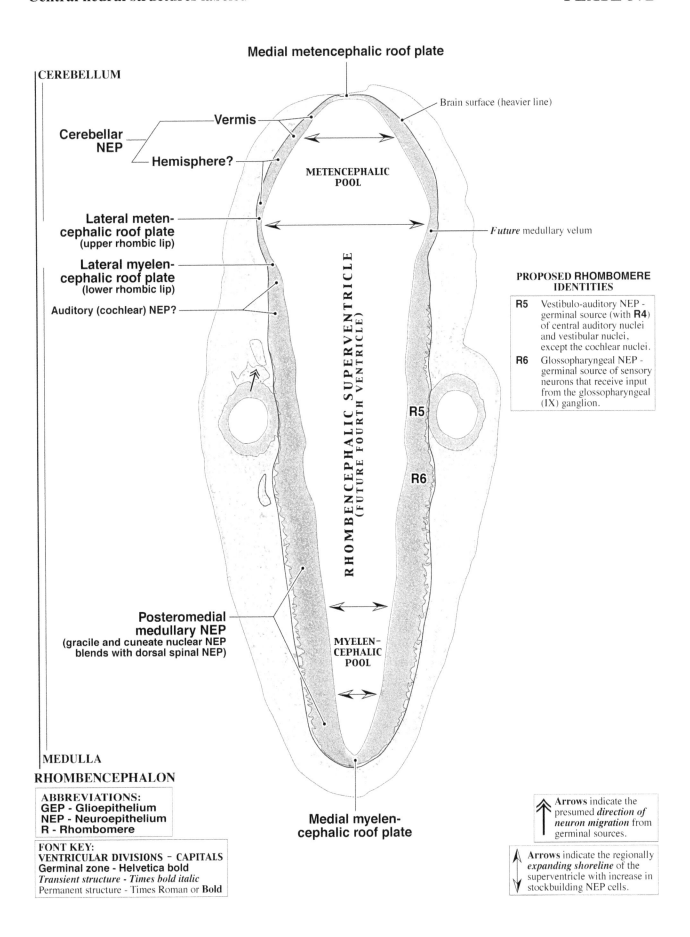

Medial metencephalic roof plate

CEREBELLUM

Vermis

Brain surface (heavier line)

Cerebellar NEP

METENCEPHALIC POOL

Hemisphere?

Lateral meten-cephalic roof plate
(upper rhombic lip)

Future medullary velum

Lateral myelen-cephalic roof plate
(lower rhombic lip)

Auditory (cochlear) NEP?

RHOMBENCEPHALIC SUPERVENTRICLE
(FUTURE FOURTH VENTRICLE)

PROPOSED RHOMBOMERE IDENTITIES

R5	Vestibulo-auditory NEP - germinal source (with **R4**) of central auditory nuclei and vestibular nuclei, except the cochlear nuclei.
R6	Glossopharyngeal NEP - germinal source of sensory neurons that receive input from the glossopharyngeal (IX) ganglion.

R5

R6

Posteromedial medullary NEP
(gracile and cuneate nuclear NEP blends with dorsal spinal NEP)

MYELEN-CEPHALIC POOL

MEDULLA

RHOMBENCEPHALON

ABBREVIATIONS:
GEP - Glioepithelium
NEP - Neuroepithelium
R - Rhombomere

Medial myelen-cephalic roof plate

FONT KEY:
VENTRICULAR DIVISIONS – CAPITALS
Germinal zone - Helvetica bold
Transient structure - Times bold italic
Permanent structure - Times Roman or **Bold**

Arrows indicate the presumed *direction of neuron migration* from germinal sources.

Arrows indicate the regionally *expanding shoreline* of the superventricle with increase in stockbuilding NEP cells.

PART V: C9297
CR 4.5 mm (GW 4.5)
Sagittal

Carnegie Collection specimen #9297 (designated here as C9297) with a 4.5-mm crown-rump length (CR) is estimated to be at gestational week (GW) 4.5. C9297 was embedded in a celloidin/paraffin mix and was cut in 8-μm sagittal sections that were stained with azan. Various orientations of the computer-aided 3-D reconstruction of C836's brain are used to show the gross external features of a GW4 brain (**Figure 9**). Like most sagittally cut specimens, C9297's sections are not parallel to the midline; **Figure 9** shows the approximate rotations from the midline in front (**B**) and back views (**C**). We photographed 18 sections at low magnification from the left to right sides of the brain. Nine of the sections, mainly from the left side of the brain, are illustrated in **Plates 38AB** to **46AB**. Each illustrated section shows the brain with all surrounding tissues. Labels in **A Plates** (normal-contrast images) identify the approximate midline, non-neural structures, peripheral neural structures, and brain ventricular divisions; labels in **B Plates** (low-contrast images) identify central neural structures. **Plates 47AB** to **51AB** show high-magnification views of the brain.

The prosencephalon is the smallest major brain structure with little distinction between a future telencephalon and diencephalon. The entire prosencephalic neuroepithelium is rapidly stockbuilding its various populations of neuronal and glial stem cells. The lamina terminalis in the ventral prosencephalon marks the site of closure of the anterior neuropore. The optic vesicle is still prominent, but no definite lens placode is present. A cell-dense area adjacent to the olfactory placode may be supporting cells associated with growth of the olfactory nerve toward the brain.

The stockbuilding pretectal and tectal neuroepithelia in the mesencephalon blend with the presumptive cerebellar neuroepithelium in the dorsomedial rhombencephalon. The stockbuilding tegmental and isthmal neuroepithelia form the mesencephalic floor. A thin subpial fiber band lines the tegmentum and isthmus. The entire mesencephalon forms the prominent mesencephalic arch, one of the most distinctive features of the brain in the early first trimester.

The rhombencephalon is the largest brain structure, with rhombomeres 2 through 7 forming well-defined swellings in the lateral neuroepithelium. These swellings have been prominent in every first trimester specimen in this Volume. There is strong anatomical evidence that sensory cranial ganglia of nerves V through X and the otic vesicle are located directly lateral to the rhombomeres with which they interact. The trigeminal ganglion (V afferents) appears in sections lateral to the last section that contains rhombomere 2. The vestibulocochlear ganglion (VIII afferents) is lateral to the last section that contains rhombomere 4; the otic vesicle is lateral to the last section that contains rhombomere 5. The presumptive superior glossopharyngeal ganglion (IX afferents) is lateral to the last section with rhombomere 6, and the large superior vagal ganglion (X afferents) is lateral to the last section with rhombomere 7. The association of rhombomere 3 with the sensory part of nerve VII is less clear, but the facial ganglion (VII afferents) is near its presumptive placodal source in lateral sections. Each rhombomere has a feathered basal edge of the NEP in most lateral sections, where the outer edges of the rhombomeric neuroepithelium are cut tangentially. A few sensory axons may have entered the brain from sensory ganglia and add to the feathering of the NEP outer edge. There are clumps of cells outside the brain that mark the entry points of cranial nerves—the boundary caps that may be Schwann cell precursors. Rhombomeres fade out as sections get closer to the midline because those cuts are mainly in basal plate NEP. No migrating cells are outside the lower medullary neuroepithelium. The subpial fiber band is thicker as the brain blends with the spinal cord. The cerebellum stands out as the most immature and smallest rhombencephalic structure. The most lateral sections show less dense cells in the basal NEP that appear to be migrating neurons, but it is still too early to see the oldest deep nuclear neurons. The feathered basal edge is an artifact of tangentially cutting the cerebellar NEP.

EXTERNAL FEATURES OF THE GW4 BRAIN

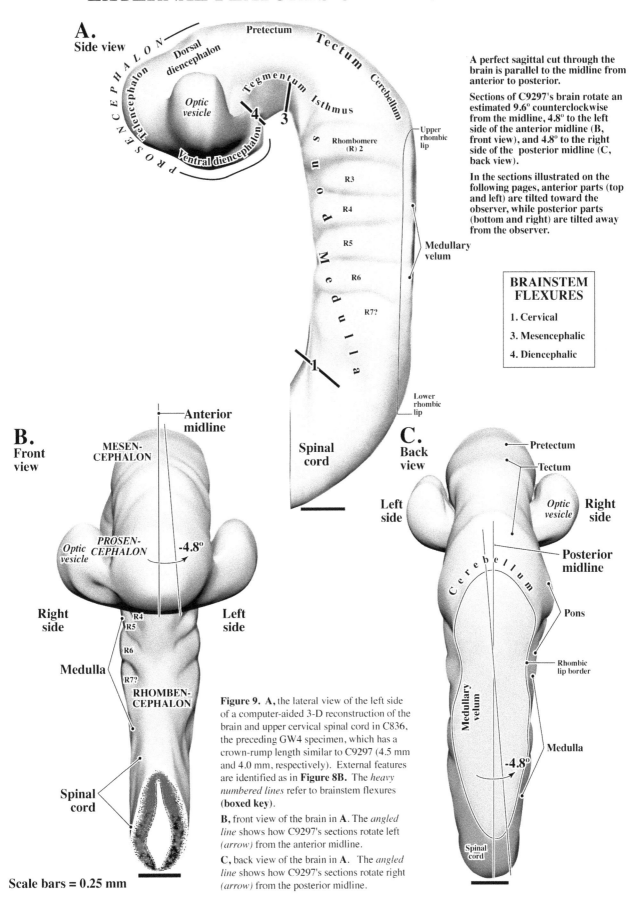

A. Side view

PROSENCEPHALON

Telencephalon

Dorsal diencephalon

Optic vesicle

Pretectum

Tectum

Tegmentum

Cerebellum

Isthmus

Ventral diencephalon

4

3

Rhombomere (R) 2

R3

R4

R5

R6

R7?

Pons

Medulla

Upper rhombic lip

Medullary velum

Lower rhombic lip

Spinal cord

A perfect sagittal cut through the brain is parallel to the midline from anterior to posterior.

Sections of C9297's brain rotate an estimated 9.6° counterclockwise from the midline, 4.8° to the left side of the anterior midline (B, front view), and 4.8° to the right side of the posterior midline (C, back view).

In the sections illustrated on the following pages, anterior parts (top and left) are tilted toward the observer, while posterior parts (bottom and right) are tilted away from the observer.

BRAINSTEM FLEXURES

1. Cervical
3. Mesencephalic
4. Diencephalic

B. Front view

MESEN-CEPHALON

PROSEN-CEPHALON

Optic vesicle

-4.8°

Anterior midline

Right side

Left side

R4
R5
R6
R7?

Medulla

RHOMBEN-CEPHALON

Spinal cord

C. Back view

Left side

Right side

Pretectum

Tectum

Optic vesicle

Posterior midline

Cerebellum

Pons

Rhombic lip border

Medullary velum

Medulla

-4.8°

Spinal cord

Figure 9. A, the lateral view of the left side of a computer-aided 3-D reconstruction of the brain and upper cervical spinal cord in C836, the preceding GW4 specimen, which has a crown-rump length similar to C9297 (4.5 mm and 4.0 mm, respectively). External features are identified as in **Figure 8B**. The *heavy numbered lines* refer to brainstem flexures **(boxed key)**.

B, front view of the brain in **A**. The *angled line* shows how C9297's sections rotate left *(arrow)* from the anterior midline.

C, back view of the brain in **A**. The *angled line* shows how C9297's sections rotate right *(arrow)* from the posterior midline.

Scale bars = 0.25 mm

PLATE 38A

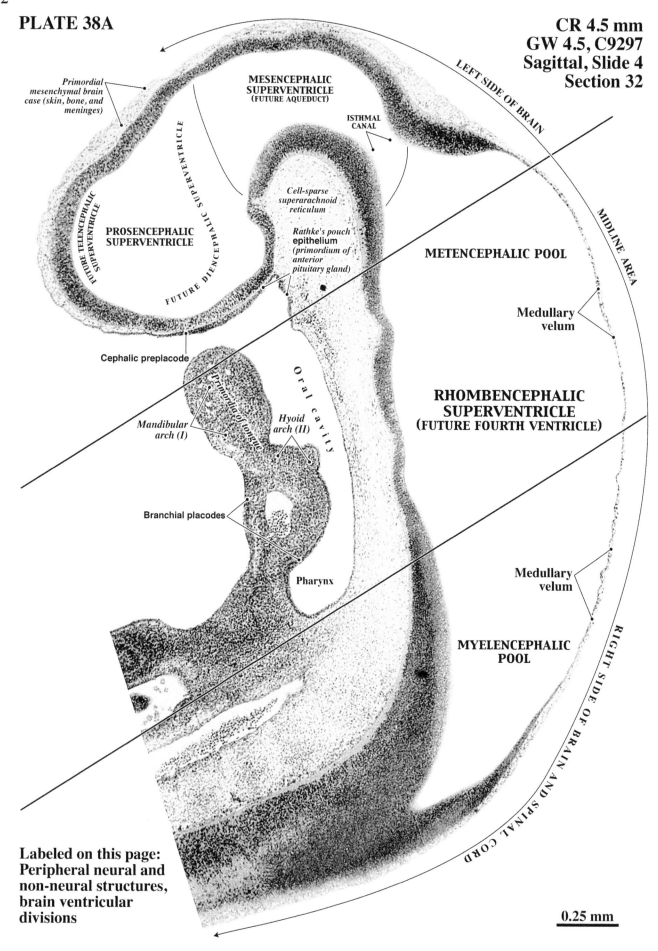

Primordial mesenchymal brain case (skin, bone, and meninges)

MESENCEPHALIC SUPERVENTRICLE
(FUTURE AQUEDUCT)

ISTHMAL CANAL

LEFT SIDE OF BRAIN

FUTURE DIENCEPHALIC SUPERVENTRICLE

FUTURE TELENCEPHALIC SUPERVENTRICLE

Cell-sparse superarachnoid reticulum

PROSENCEPHALIC SUPERVENTRICLE

Rathke's pouch epithelium (*primordium of anterior pituitary gland*)

METENCEPHALIC POOL

MIDLINE AREA

Medullary velum

Cephalic preplacode

Primordium of tongue

Oral cavity

Mandibular arch (I)

Hyoid arch (II)

RHOMBENCEPHALIC SUPERVENTRICLE
(FUTURE FOURTH VENTRICLE)

Branchial placodes

Pharynx

Medullary velum

MYELENCEPHALIC POOL

RIGHT SIDE OF BRAIN AND SPINAL CORD

**Labeled on this page:
Peripheral neural and
non-neural structures,
brain ventricular
divisions**

0.25 mm

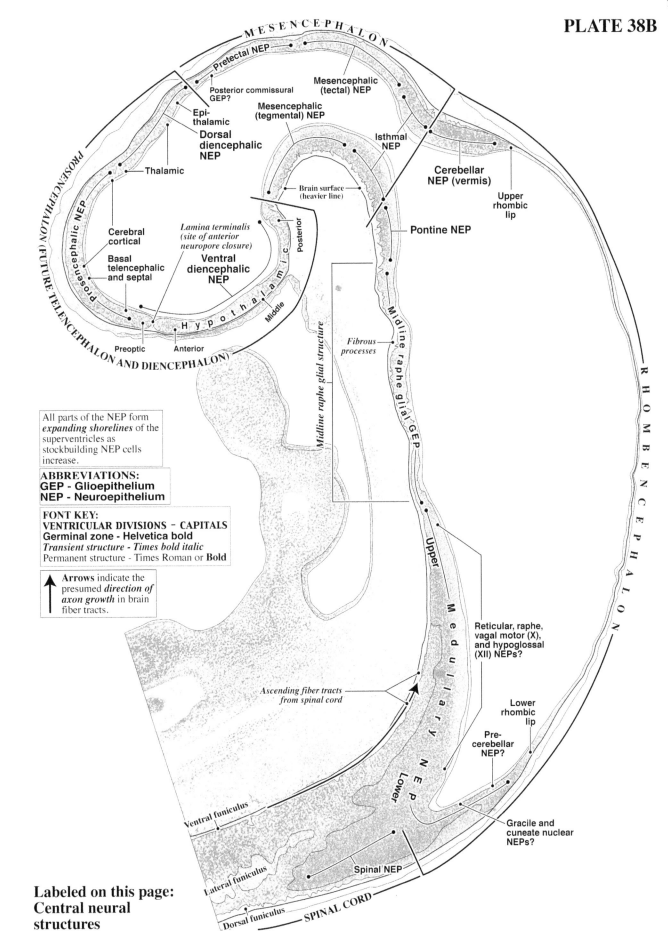

MESENCEPHALON

Pretectal NEP

Posterior commissural
GEP?

Mesencephalic
(tectal) NEP

Epi-
thalamic

Mesencephalic
(tegmental) NEP

Isthmal
NEP

**Dorsal
diencephalic
NEP**

Thalamic

Cerebellar
NEP (vermis)

Upper
rhombic
lip

PROSENCEPHALON (FUTURE TELENCEPHALON AND DIENCEPHALON)

Prosencephalic NEP

Brain surface
(heavier line)

Pontine NEP

Cerebral
cortical

*Lamina terminalis
(site of anterior
neuropore closure)*

Basal
telencephalic
and septal

**Ventral
diencephalic
NEP**

Posterior

H y p o t h a l a m i c

Middle

Preoptic

Anterior

Midline raphe glial structure

*Fibrous
processes*

Midline raphe glial GEP

R H O M B E N C E P H A L O N

All parts of the NEP form
expanding shorelines of the
superventricles as
stockbuilding NEP cells
increase.

**ABBREVIATIONS:
GEP - Glioepithelium
NEP - Neuroepithelium**

**FONT KEY:
VENTRICULAR DIVISIONS – CAPITALS
Germinal zone - Helvetica bold
Transient structure - Times bold italic
Permanent structure - Times Roman or Bold**

Arrows indicate the
presumed *direction of
axon growth* in brain
fiber tracts.

Upper

M e d u l l a r y   N E P

Reticular, raphe,
vagal motor (X),
and hypoglossal
(XII) NEPs?

*Ascending fiber tracts
from spinal cord*

Lower
rhombic
lip

Pre-
cerebellar
NEP?

Lower

Gracile and
cuneate nuclear
NEPs?

Ventral funiculus

Lateral funiculus

Spinal NEP

**Labeled on this page:
Central neural
structures**

Dorsal funiculus

SPINAL CORD

94

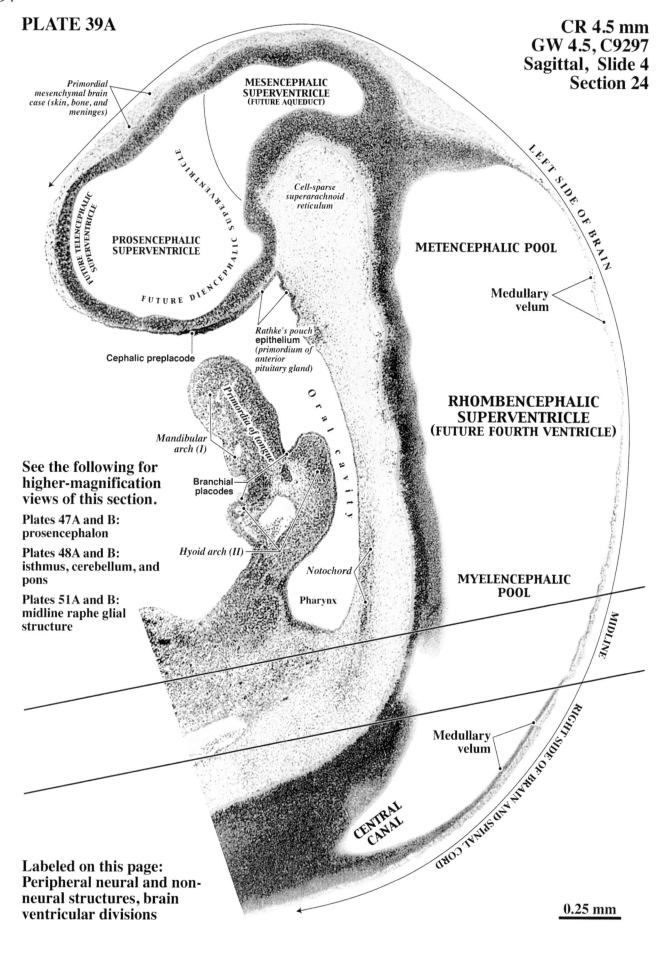

Primordial mesenchymal brain case (skin, bone, and meninges)

MESENCEPHALIC SUPERVENTRICLE (FUTURE AQUEDUCT)

LEFT SIDE OF BRAIN

FUTURE MESENCEPHALIC SUPERVENTRICLE

Cell-sparse superarachnoid reticulum

FUTURE TELENCEPHALIC SUPERVENTRICLE

PROSENCEPHALIC SUPERVENTRICLE

METENCEPHALIC POOL

FUTURE DIENCEPHALIC SUPERVENTRICLE

Medullary velum

Cephalic preplacode

Rathke's pouch epithelium *(primordium of anterior pituitary gland)*

O r a l   c a v i t y

RHOMBENCEPHALIC SUPERVENTRICLE (FUTURE FOURTH VENTRICLE)

Primordia of tongue

Mandibular arch (I)

Branchial placodes

See the following for higher-magnification views of this section.

Plates 47A and B: prosencephalon

Plates 48A and B: isthmus, cerebellum, and pons

Plates 51A and B: midline raphe glial structure

Hyoid arch (II)

Notochord

MYELENCEPHALIC POOL

MIDLINE

Pharynx

Medullary velum

RIGHT SIDE OF BRAIN AND SPINAL CORD

CENTRAL CANAL

Labeled on this page: Peripheral neural and non-neural structures, brain ventricular divisions

0.25 mm

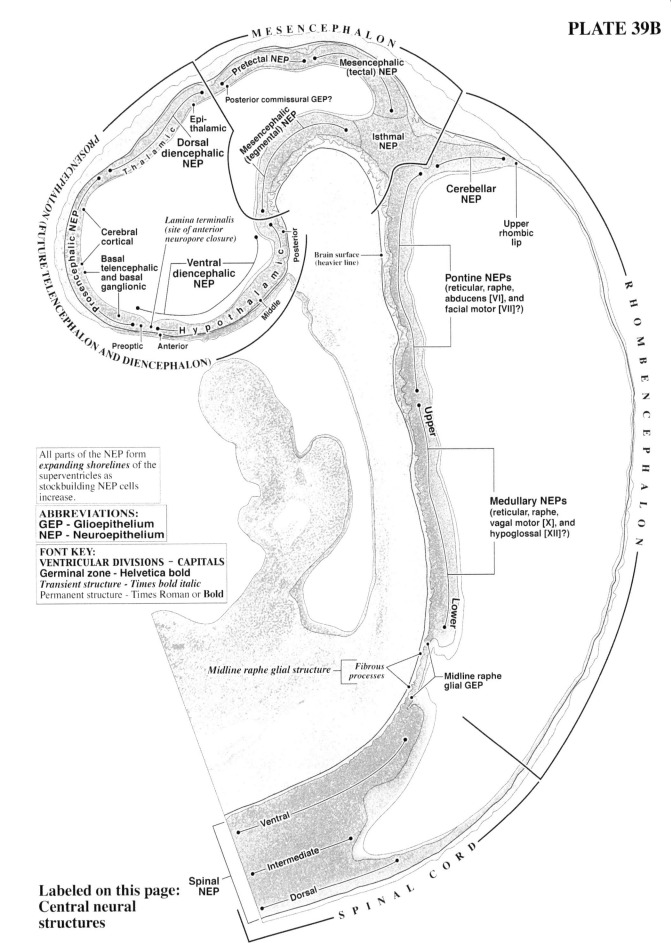

MESENCEPHALON

Pretectal NEP

Mesencephalic (tectal) NEP

Posterior commissural GEP?

Epi-thalamic

Dorsal diencephalic NEP

Mesencephalic (tegmental) NEP

Isthmal NEP

Cerebellar NEP

Upper rhombic lip

T=h=a=l=a=m=i=c

Cerebral cortical

Lamina terminalis (site of anterior neuropore closure)

Posterior

Brain surface (heavier line)

Basal telencephalic and basal ganglionic

Ventral diencephalic NEP

Middle

Pontine NEPs (reticular, raphe, abducens [VI], and facial motor [VII]?)

Preoptic   Anterior

H y p o t h a l a m i c

PROSENCEPHALON

(FUTURE TELENCEPHALON AND DIENCEPHALON)

Prosencephalic NEP

RHOMBENCEPHALON

Upper

Medullary NEPs (reticular, raphe, vagal motor [X], and hypoglossal [XII]?)

Lower

All parts of the NEP form *expanding shorelines* of the superventricles as stockbuilding NEP cells increase.

ABBREVIATIONS:
GEP - Glioepithelium
NEP - Neuroepithelium

FONT KEY:
VENTRICULAR DIVISIONS – CAPITALS
Germinal zone - Helvetica bold
Transient structure - Times bold italic
Permanent structure - Times Roman or **Bold**

Midline raphe glial structure

Fibrous processes

Midline raphe glial GEP

Ventral

Intermediate

Spinal NEP

Dorsal

SPINAL CORD

Labeled on this page: Central neural structures

96

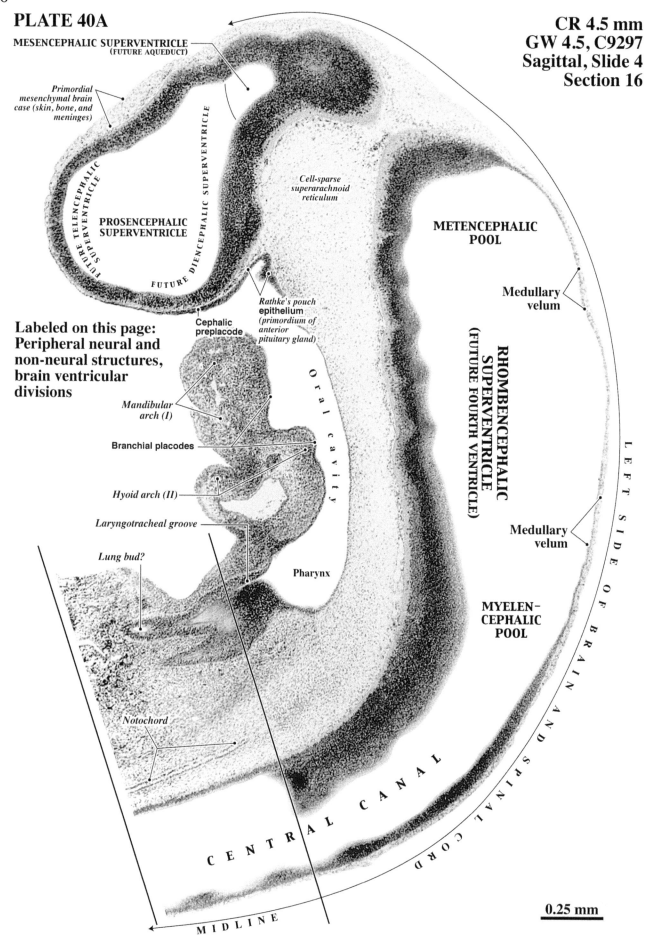

PLATE 40A

MESENCEPHALIC SUPERVENTRICLE
(FUTURE AQUEDUCT)

CR 4.5 mm
GW 4.5, C9297
Sagittal, Slide 4
Section 16

Primordial mesenchymal brain case (skin, bone, and meninges)

Cell-sparse superarachnoid reticulum

FUTURE TELENCEPHALIC SUPERVENTRICLE

PROSENCEPHALIC SUPERVENTRICLE

FUTURE DIENCEPHALIC SUPERVENTRICLE

METENCEPHALIC POOL

Medullary velum

Labeled on this page:
Peripheral neural and
non-neural structures,
brain ventricular
divisions

Cephalic preplacode

Rathke's pouch epithelium *(primordium of anterior pituitary gland)*

O r a l   c a v i t y

RHOMBENCEPHALIC SUPERVENTRICLE
(FUTURE FOURTH VENTRICLE)

Mandibular arch (I)

Branchial placodes

Hyoid arch (II)

Medullary velum

Laryngotracheal groove

Lung bud?

Pharynx

MYELEN- CEPHALIC POOL

L E F T   S I D E   O F   B R A I N   A N D   S P I N A L   C O R D

Notochord

C E N T R A L   C A N A L

0.25 mm

M I D L I N E

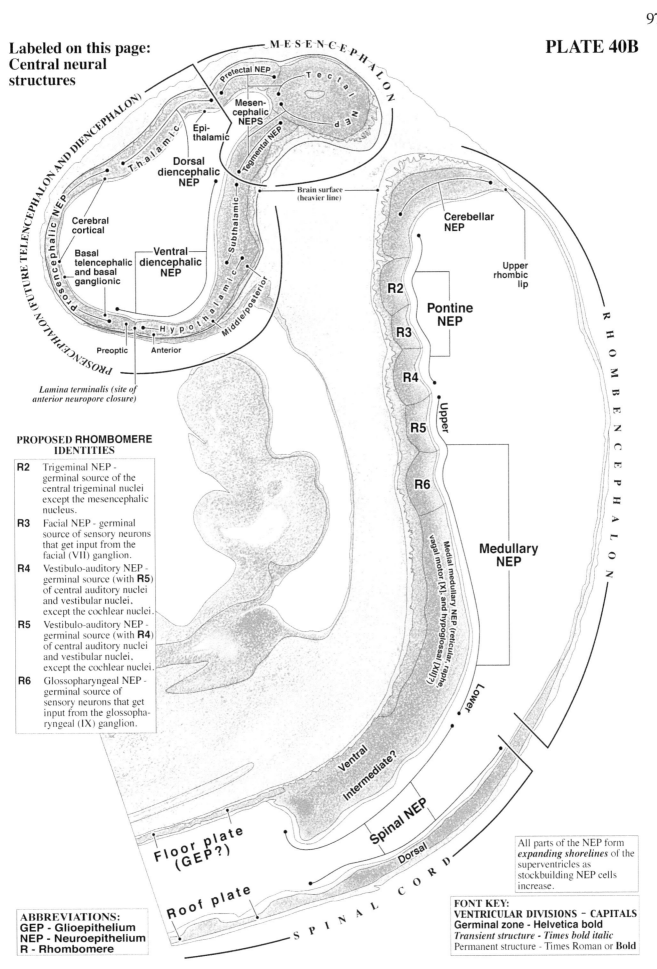

Labeled on this page:
Central neural
structures

PROPOSED RHOMBOMERE
IDENTITIES

R2 — Trigeminal NEP - germinal source of the central trigeminal nuclei except the mesencephalic nucleus.

R3 — Facial NEP - germinal source of sensory neurons that get input from the facial (VII) ganglion.

R4 — Vestibulo-auditory NEP - germinal source (with R5) of central auditory nuclei and vestibular nuclei, except the cochlear nuclei.

R5 — Vestibulo-auditory NEP - germinal source (with R4) of central auditory nuclei and vestibular nuclei, except the cochlear nuclei.

R6 — Glossopharyngeal NEP - germinal source of sensory neurons that get input from the glossopharyngeal (IX) ganglion.

All parts of the NEP form *expanding shorelines* of the superventricles as stockbuilding NEP cells increase.

ABBREVIATIONS:
GEP - Glioepithelium
NEP - Neuroepithelium
R - Rhombomere

FONT KEY:
VENTRICULAR DIVISIONS – CAPITALS
Germinal zone - Helvetica bold
Transient structure - Times bold italic
Permanent structure - Times Roman or **Bold**

98

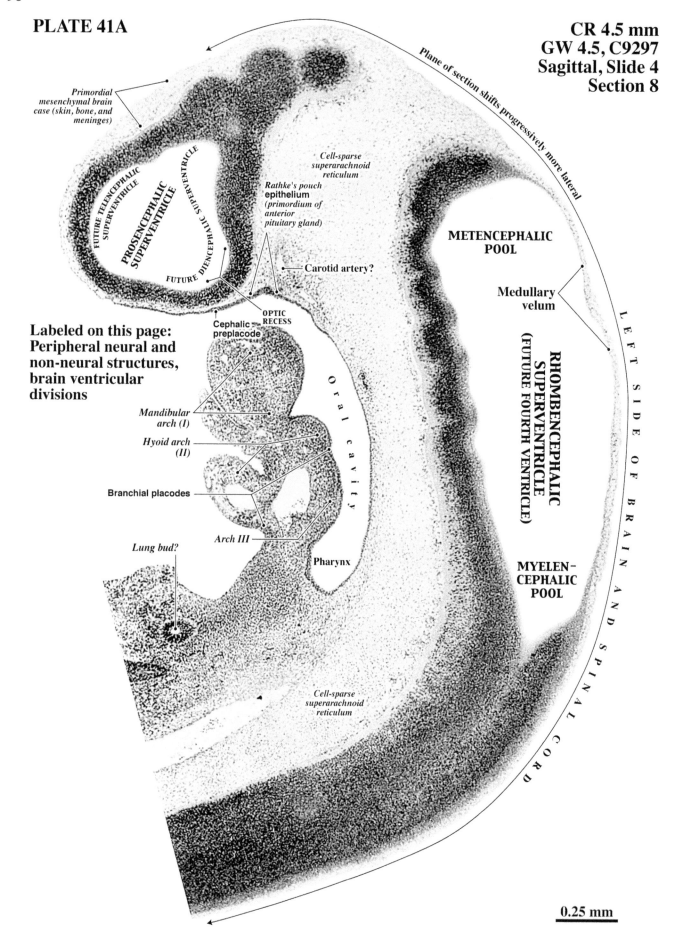

Primordial mesenchymal brain case (skin, bone, and meninges)

Plane of section shifts progressively more lateral

FUTURE TELENCEPHALIC SUPERVENTRICLE

PROSENCEPHALIC SUPERVENTRICLE

FUTURE DIENCEPHALIC SUPERVENTRICLE

Cell-sparse superarachnoid reticulum

Rathke's pouch epithelium *(primordium of anterior pituitary gland)*

METENCEPHALIC POOL

Medullary velum

Carotid artery?

FUTURE

OPTIC RECESS

Cephalic preplacode

Labeled on this page: Peripheral neural and non-neural structures, brain ventricular divisions

O r a l   c a v i t y

RHOMBENCEPHALIC SUPERVENTRICLE (FUTURE FOURTH VENTRICLE)

L E F T   S I D E   O F   B R A I N   A N D   S P I N A L   C O R D

Mandibular arch (I)

Hyoid arch (II)

Branchial placodes

Arch III

Lung bud?

Pharynx

MYELEN- CEPHALIC POOL

Cell-sparse superarachnoid reticulum

0.25 mm

Labeled on this page: Central neural structures

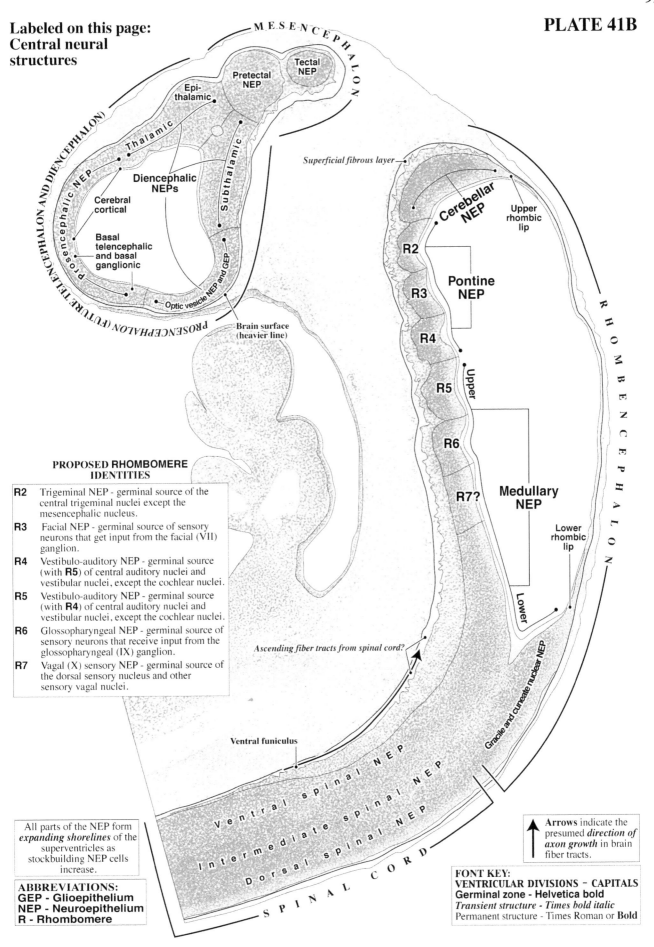

PROPOSED RHOMBOMERE IDENTITIES

R2 Trigeminal NEP - germinal source of the central trigeminal nuclei except the mesencephalic nucleus.

R3 Facial NEP - germinal source of sensory neurons that get input from the facial (VII) ganglion.

R4 Vestibulo-auditory NEP - germinal source (with **R5**) of central auditory nuclei and vestibular nuclei, except the cochlear nuclei.

R5 Vestibulo-auditory NEP - germinal source (with **R4**) of central auditory nuclei and vestibular nuclei, except the cochlear nuclei.

R6 Glossopharyngeal NEP - germinal source of sensory neurons that receive input from the glossopharyngeal (IX) ganglion.

R7 Vagal (X) sensory NEP - germinal source of the dorsal sensory nucleus and other sensory vagal nuclei.

All parts of the NEP form *expanding shorelines* of the superventricles as stockbuilding NEP cells increase.

ABBREVIATIONS:
GEP - Glioepithelium
NEP - Neuroepithelium
R - Rhombomere

Arrows indicate the presumed *direction of axon growth* in brain fiber tracts.

FONT KEY:
VENTRICULAR DIVISIONS - CAPITALS
Germinal zone - Helvetica bold
Transient structure - Times bold italic
Permanent structure - Times Roman or **Bold**

100

Labeled on this page:
Peripheral neural and
non-neural structures,
brain ventricular
divisions

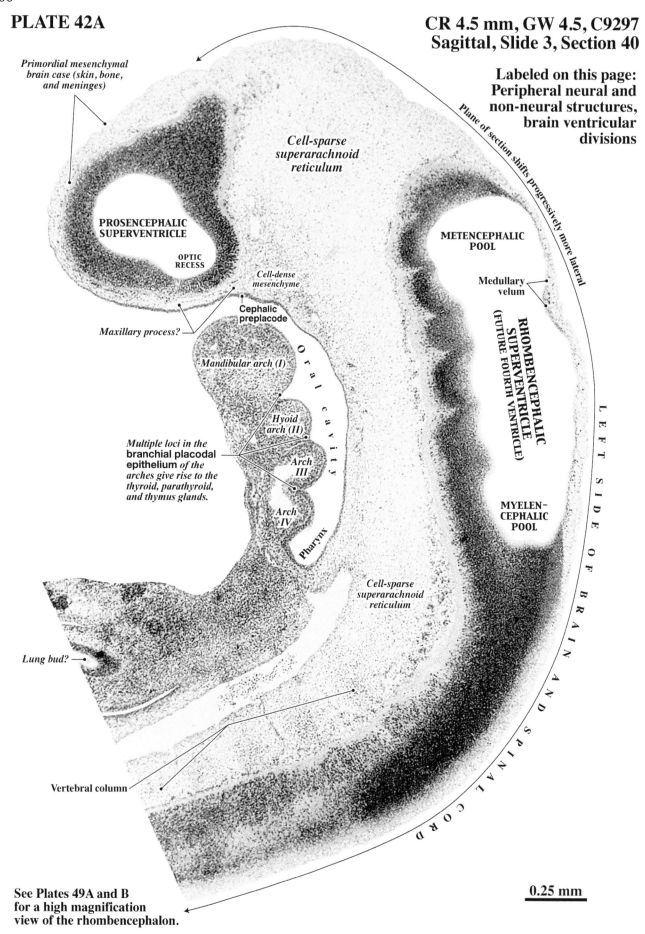

*Primordial mesenchymal
brain case (skin, bone,
and meninges)*

*Cell-sparse
superarachnoid
reticulum*

Plane of section shifts progressively more lateral

**PROSENCEPHALIC
SUPERVENTRICLE**

**OPTIC
RECESS**

**METENCEPHALIC
POOL**

Medullary
velum

*Cell-dense
mesenchyme*

Cephalic
preplacode

Maxillary process?

Mandibular arch (I)

O r a l   c a v i t y

**RHOMBENCEPHALIC
SUPERVENTRICLE
(FUTURE FOURTH VENTRICLE)**

*Hyoid
arch (II)*

Multiple loci in the
**branchial placodal
epithelium** *of the
arches give rise to the
thyroid, parathyroid,
and thymus glands.*

*Arch
III*

**MYELEN-
CEPHALIC
POOL**

*Arch
IV*

Pharynx

*Cell-sparse
superarachnoid
reticulum*

L E F T   S I D E   O F   B R A I N   A N D   S P I N A L   C O R D

Lung bud?

Vertebral column

See Plates 49A and B
for a high magnification
view of the rhombencephalon.

0.25 mm

Labeled on this page:
Central neural
structures

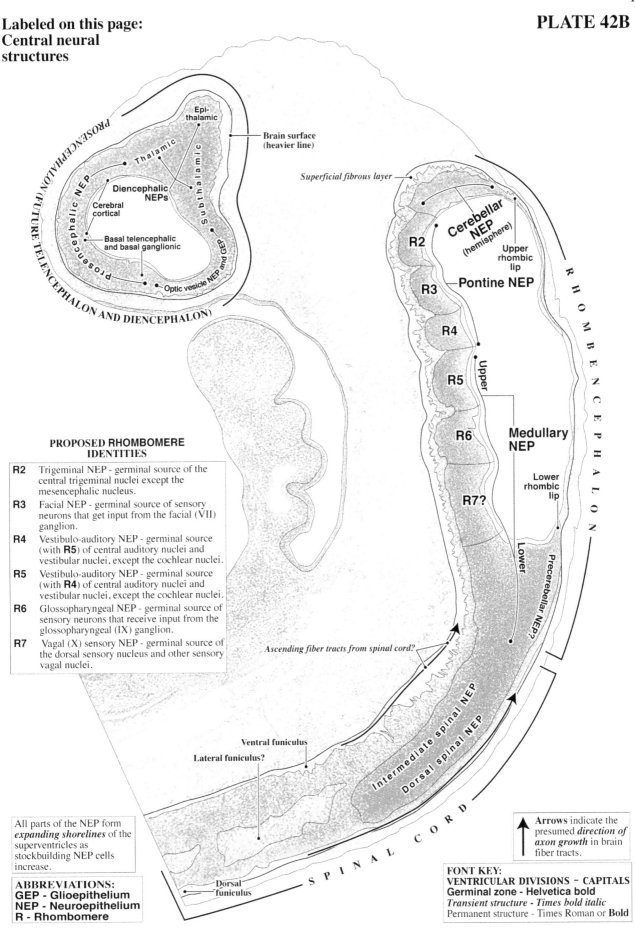

PROPOSED RHOMBOMERE
IDENTITIES

R2   Trigeminal NEP - germinal source of the
      central trigeminal nuclei except the
      mesencephalic nucleus.
R3   Facial NEP - germinal source of sensory
      neurons that get input from the facial (VII)
      ganglion.
R4   Vestibulo-auditory NEP - germinal source
      (with R5) of central auditory nuclei and
      vestibular nuclei, except the cochlear nuclei.
R5   Vestibulo-auditory NEP - germinal source
      (with R4) of central auditory nuclei and
      vestibular nuclei, except the cochlear nuclei.
R6   Glossopharyngeal NEP - germinal source of
      sensory neurons that receive input from the
      glossopharyngeal (IX) ganglion.
R7   Vagal (X) sensory NEP - germinal source of
      the dorsal sensory nucleus and other sensory
      vagal nuclei.

All parts of the NEP form
expanding shorelines of the
superventricles as
stockbuilding NEP cells
increase.

ABBREVIATIONS:
GEP - Glioepithelium
NEP - Neuroepithelium
R - Rhombomere

Arrows indicate the
presumed *direction of
axon growth* in brain
fiber tracts.

FONT KEY:
VENTRICULAR DIVISIONS – CAPITALS
Germinal zone - Helvetica bold
Transient structure - Times bold italic
Permanent structure - Times Roman or **Bold**

CR 4.5 mm, GW 4.5, C9297
Sagittal, Slide 3, Section 32

Labeled on this page:
Peripheral neural and
non-neural structures,
brain ventricular
divisions

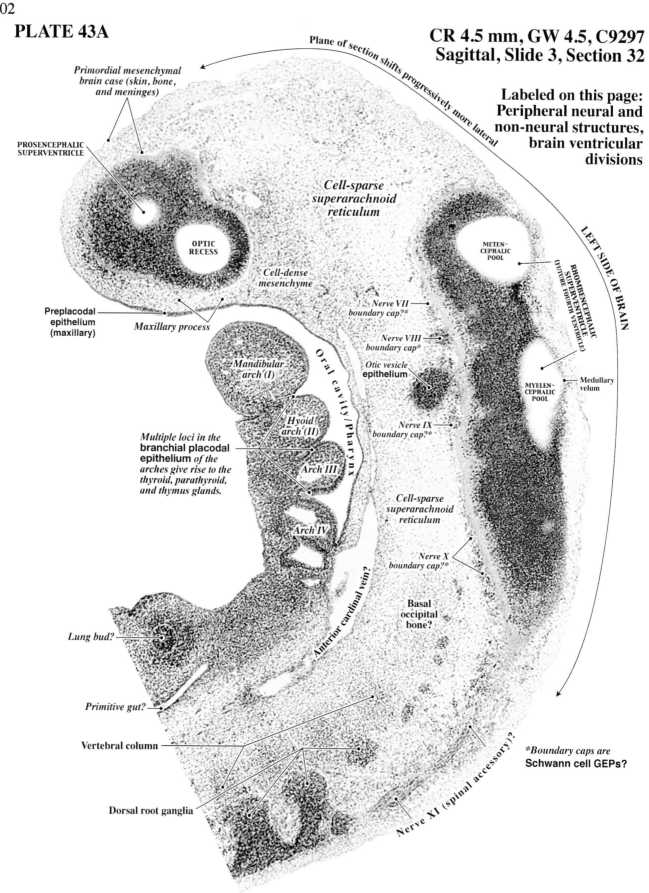

Plane of section shifts progressively more lateral

*Primordial mesenchymal
brain case (skin, bone,
and meninges)*

PROSENCEPHALIC
SUPERVENTRICLE

*Cell-sparse
superarachnoid
reticulum*

METEN-
CEPHALIC
POOL

LEFT SIDE OF BRAIN

RHOMBENCEPHALIC
SUPERVENTRICLE
(FUTURE FOURTH VENTRICLE)

OPTIC
RECESS

*Cell-dense
mesenchyme*

*Nerve VII
boundary cap?**

Preplacodal
epithelium
(maxillary)

Maxillary process

*Nerve VIII
boundary cap**

Otic vesicle
epithelium

MYELEN-
CEPHALIC
POOL

Medullary
velum

*Mandibular
arch (I)*

Oral cavity/Pharynx

*Nerve IX
boundary cap?**

*Hyoid
arch (II)*

Multiple loci in the
**branchial placodal
epithelium** *of the
arches give rise to the
thyroid, parathyroid,
and thymus glands.*

Arch III

Arch IV

*Cell-sparse
superarachnoid
reticulum*

*Nerve X
boundary cap?**

Anterior cardinal vein?

**Basal
occipital
bone?**

Lung bud?

Primitive gut?

Vertebral column

Nerve XI (spinal accessory)?

*Boundary caps are
Schwann cell GEPs?

Dorsal root ganglia

0.25 mm

Labeled on this page:
Central neural
structures

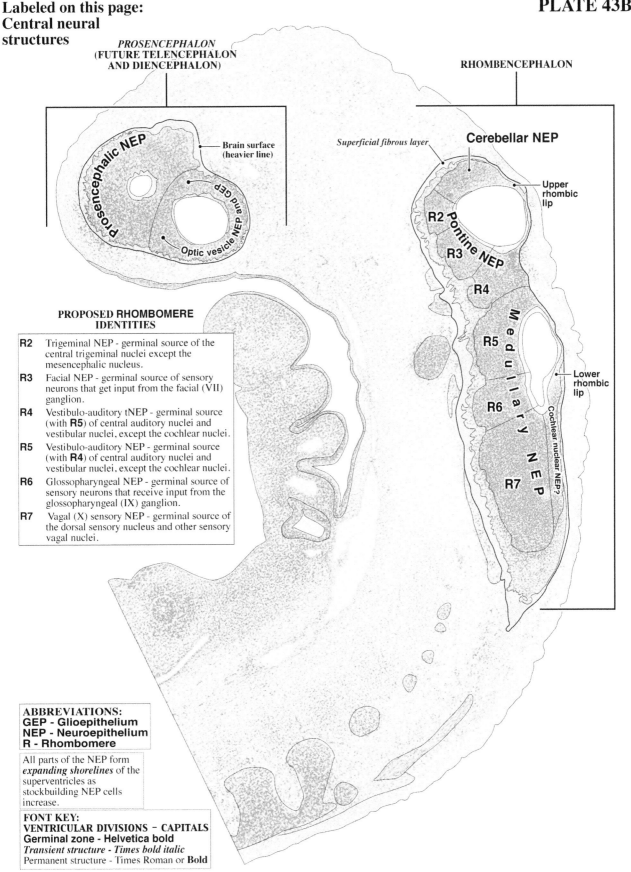

PROSENCEPHALON
(FUTURE TELENCEPHALON
AND DIENCEPHALON)

RHOMBENCEPHALON

Prosencephalic NEP

Brain surface
(heavier line)

Superficial fibrous layer

Cerebellar NEP

Upper
rhombic
lip

R2

Pontine NEP

R3

R4

Optic vesicle NEP and GEP

R5

Medullary NEP

Lower
rhombic
lip

R6

Cochlear nuclear NEP?

R7

PROPOSED RHOMBOMERE
IDENTITIES

R2   Trigeminal NEP - germinal source of the central trigeminal nuclei except the mesencephalic nucleus.

R3   Facial NEP - germinal source of sensory neurons that get input from the facial (VII) ganglion.

R4   Vestibulo-auditory tNEP - germinal source (with **R5**) of central auditory nuclei and vestibular nuclei, except the cochlear nuclei.

R5   Vestibulo-auditory NEP - germinal source (with **R4**) of central auditory nuclei and vestibular nuclei, except the cochlear nuclei.

R6   Glossopharyngeal NEP - germinal source of sensory neurons that receive input from the glossopharyngeal (IX) ganglion.

R7   Vagal (X) sensory NEP - germinal source of the dorsal sensory nucleus and other sensory vagal nuclei.

ABBREVIATIONS:
GEP - Glioepithelium
NEP - Neuroepithelium
R - Rhombomere

All parts of the NEP form
expanding shorelines of the
superventricles as
stockbuilding NEP cells
increase.

FONT KEY:
VENTRICULAR DIVISIONS – CAPITALS
Germinal zone - Helvetica bold
Transient structure - Times bold italic
Permanent structure - Times Roman or **Bold**

**Labeled on this page:
Peripheral neural and
non-neural structures,
brain ventricular
divisions**

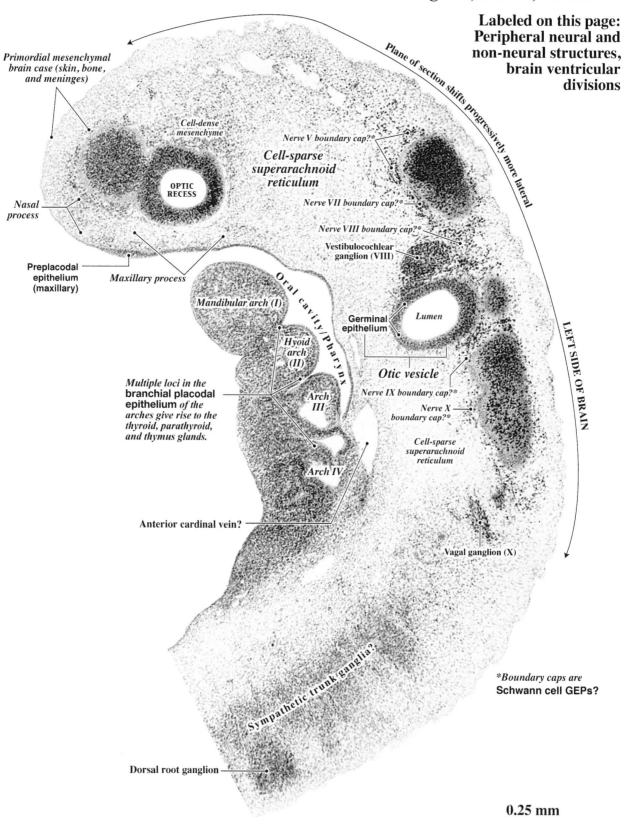

*Primordial mesenchymal
brain case (skin, bone,
and meninges)*

*Cell-dense
mesenchyme*

Nerve V boundary cap?*

*Cell-sparse
superarachnoid
reticulum*

**OPTIC
RECESS**

Nerve VII boundary cap?*

**Nasal
process**

Nerve VIII boundary cap?*

**Vestibulocochlear
ganglion (VIII)**

**Preplacodal
epithelium
(maxillary)**

Maxillary process

Oral cavity/Pharynx

**Germinal
epithelium**

Lumen

Mandibular arch (I)

*Hyoid
arch
(II)*

*Multiple loci in the
branchial placodal
epithelium of the
arches give rise to the
thyroid, parathyroid,
and thymus glands.*

*Arch
III*

Otic vesicle

Nerve IX boundary cap?*

*Nerve X
boundary cap?**

*Cell-sparse
superarachnoid
reticulum*

Arch IV

Anterior cardinal vein?

Vagal ganglion (X)

Plane of section shifts progressively more lateral

LEFT SIDE OF BRAIN

**Boundary caps are
Schwann cell GEPs?*

Sympathetic trunk ganglia?

Dorsal root ganglion

0.25 mm

**See a higher magnification
view of the rhombencephalon
in Plates 50A and B.**

Labeled on this page:
Central neural
structures

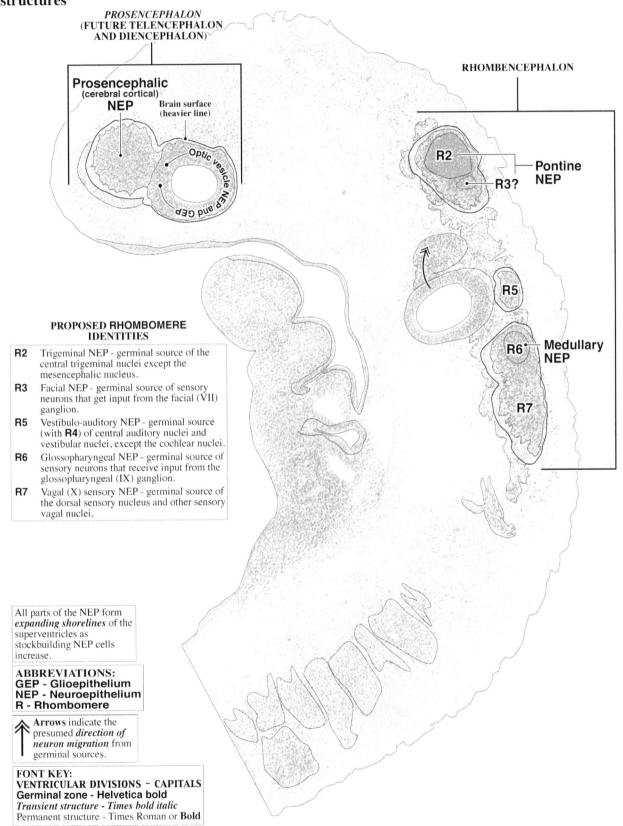

PROSENCEPHALON
(FUTURE TELENCEPHALON
AND DIENCEPHALON)

Prosencephalic
(cerebral cortical)
NEP

Brain surface
(heavier line)

Optic vesicle NEP and GEP

RHOMBENCEPHALON

R2

R3? — Pontine NEP

R5

R6 — Medullary NEP

R7

PROPOSED RHOMBOMERE
IDENTITIES

R2 Trigeminal NEP - germinal source of the
central trigeminal nuclei except the
mesencephalic nucleus.

R3 Facial NEP - germinal source of sensory
neurons that get input from the facial (VII)
ganglion.

R5 Vestibulo-auditory NEP - germinal source
(with **R4**) of central auditory nuclei and
vestibular nuclei, except the cochlear nuclei.

R6 Glossopharyngeal NEP - germinal source of
sensory neurons that receive input from the
glossopharyngeal (IX) ganglion.

R7 Vagal (X) sensory NEP - germinal source of
the dorsal sensory nucleus and other sensory
vagal nuclei.

All parts of the NEP form
expanding shorelines of the
superventricles as
stockbuilding NEP cells
increase.

ABBREVIATIONS:
GEP - Glioepithelium
NEP - Neuroepithelium
R - Rhombomere

Arrows indicate the
presumed *direction of*
neuron migration from
germinal sources.

FONT KEY:
VENTRICULAR DIVISIONS − CAPITALS
Germinal zone - Helvetica bold
Transient structure - Times bold italic
Permanent structure - Times Roman or **Bold**

106

PLATE 45A

Labeled on this page:
Non-neural structures,
brain ventricular
divisions

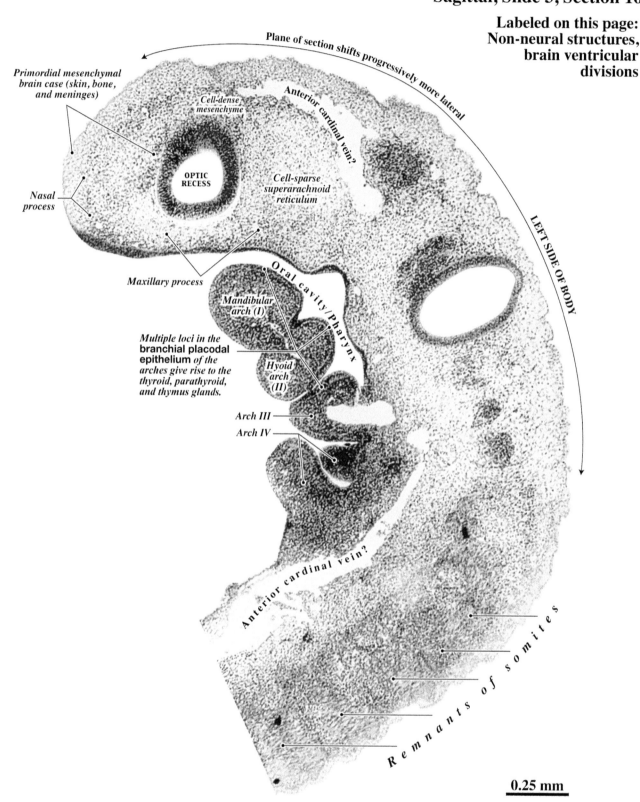

Plane of section shifts progressively more lateral

Primordial mesenchymal
brain case (skin, bone,
and meninges)

Cell-dense
mesenchyme

Anterior cardinal vein?

OPTIC
RECESS

Cell-sparse
superarachnoid
reticulum

LEFT SIDE OF BODY

Nasal
process

Maxillary process

Oral cavity/pharynx

Mandibular
arch (I)

Multiple loci in the
branchial placodal
epithelium of the
arches give rise to the
thyroid, parathyroid,
and thymus glands.

Hyoid
arch
(II)

Arch III

Arch IV

Anterior cardinal vein?

Remnants of somites

0.25 mm

Labeled on this page:
Peripheral neural
structures

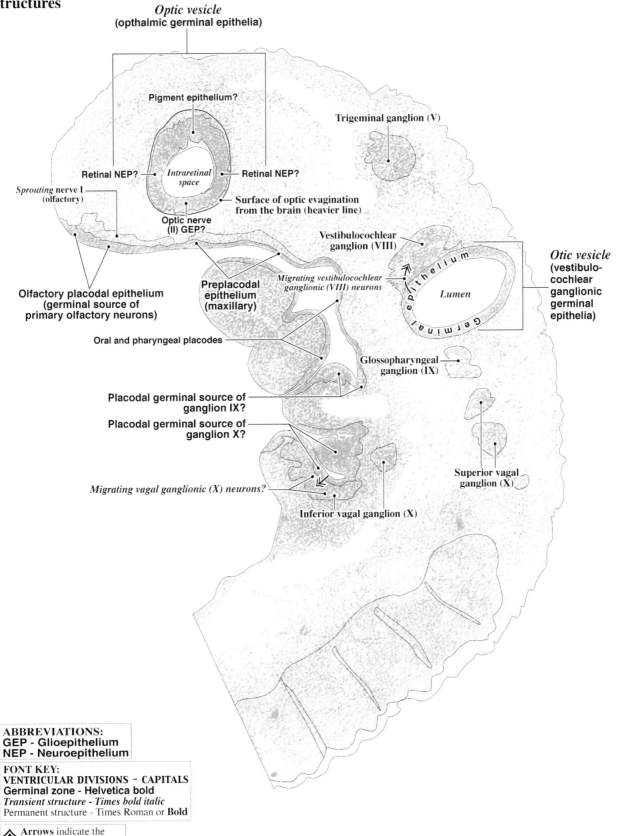

Optic vesicle
(opthalmic germinal epithelia)

Pigment epithelium?

Trigeminal ganglion (V)

Retinal NEP?

Intraretinal space

Retinal NEP?

Sprouting nerve I
(olfactory)

Surface of optic evagination
from the brain (heavier line)

Optic nerve
(II) GEP?

Vestibulocochlear
ganglion (VIII)

Otic vesicle
(vestibulo-
cochlear
ganglionic
germinal
epithelia)

Olfactory placodal epithelium
(germinal source of
primary olfactory neurons)

Preplacodal
epithelium
(maxillary)

Migrating vestibulocochlear
ganglionic (VIII) neurons

Germinal epithelium

Lumen

Oral and pharyngeal placodes

Glossopharyngeal
ganglion (IX)

Placodal germinal source of
ganglion IX?

Placodal germinal source of
ganglion X?

Migrating vagal ganglionic (X) neurons?

Superior vagal
ganglion (X)

Inferior vagal ganglion (X)

ABBREVIATIONS:
GEP - Glioepithelium
NEP - Neuroepithelium

FONT KEY:
VENTRICULAR DIVISIONS – CAPITALS
Germinal zone - Helvetica bold
Transient structure - Times bold italic
Permanent structure - Times Roman or **Bold**

Arrows indicate the
presumed *direction of*
neuron migration from
germinal sources.

PLATE 46A

CR4.5 mm, GW 4.5, C9297
Sagittal, Slide 3, Section 8

Labeled on this page:
Non-neural structures,
brain ventricular
divisions

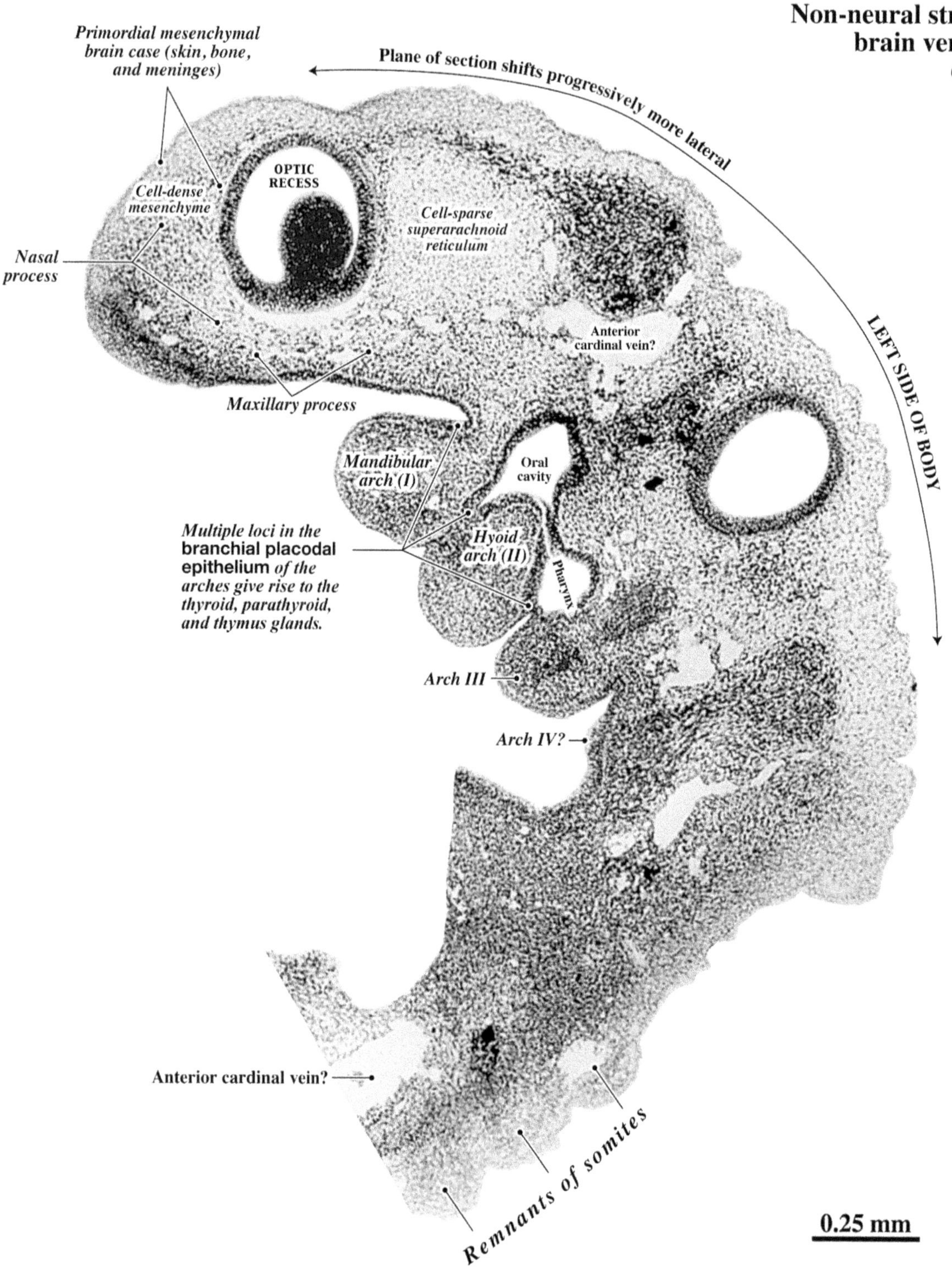

Primordial mesenchymal brain case (skin, bone, and meninges)
Plane of section shifts progressively more lateral
OPTIC RECESS
Cell-dense mesenchyme
Cell-sparse superarachnoid reticulum
Nasal process
Anterior cardinal vein?
Maxillary process
LEFT SIDE OF BODY
Mandibular arch (I)
Oral cavity
Multiple loci in the branchial placodal epithelium of the arches give rise to the thyroid, parathyroid, and thymus glands.
Hyoid arch (II)
Pharynx
Arch III
Arch IV?
Anterior cardinal vein?
Remnants of somites
0.25 mm

**Labeled on this page:
Peripheral neural
structures**

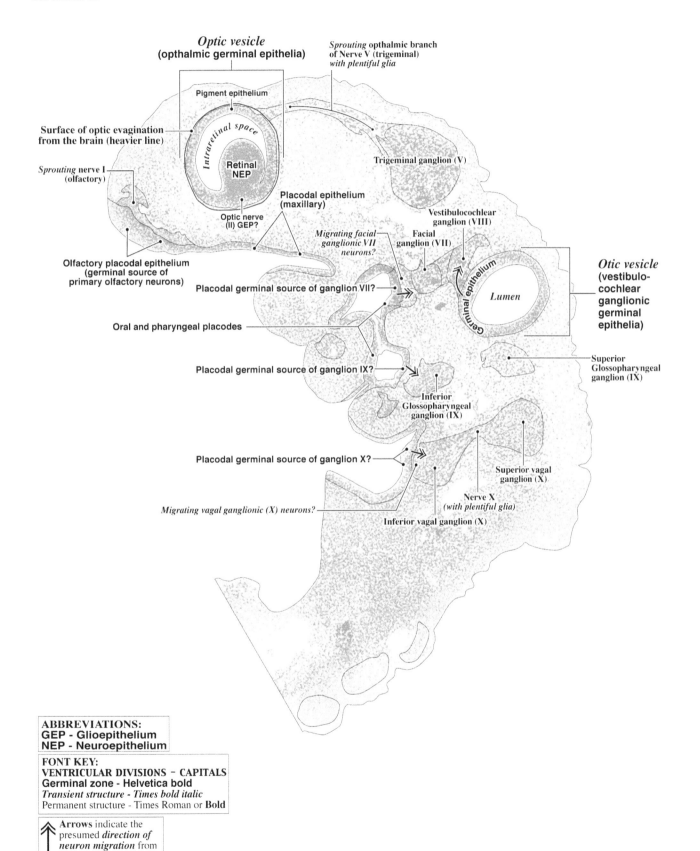

Optic vesicle
(opthalmic germinal epithelia)

Sprouting opthalmic branch
of Nerve V (trigeminal)
with plentiful glia

Pigment epithelium

Surface of optic evagination
from the brain (heavier line)

Intraretinal space

**Retinal
NEP**

Trigeminal ganglion (V)

Sprouting nerve I
(olfactory)

**Placodal epithelium
(maxillary)**

Optic nerve
(II) GEP?

*Migrating facial
ganglionic VII
neurons?*

Facial
ganglion (VII)

Vestibulocochlear
ganglion (VIII)

Olfactory placodal epithelium
(germinal source of
primary olfactory neurons)

Placodal germinal source of ganglion VII?

Germinal epithelium

Lumen

Otic vesicle
(vestibulo-
cochlear
ganglionic
germinal
epithelia)

Oral and pharyngeal placodes

Placodal germinal source of ganglion IX?

Superior
Glossopharyngeal
ganglion (IX)

Inferior
Glossopharyngeal
ganglion (IX)

Placodal germinal source of ganglion X?

Superior vagal
ganglion (X)

Migrating vagal ganglionic (X) neurons?

Nerve X
(with plentiful glia)

Inferior vagal ganglion (X)

**ABBREVIATIONS:
GEP - Glioepithelium
NEP - Neuroepithelium**

FONT KEY:
VENTRICULAR DIVISIONS - CAPITALS
Germinal zone - Helvetica bold
Transient structure - Times bold italic
Permanent structure - Times Roman or **Bold**

Arrows indicate the
presumed *direction of
neuron migration* from
germinal sources.

110

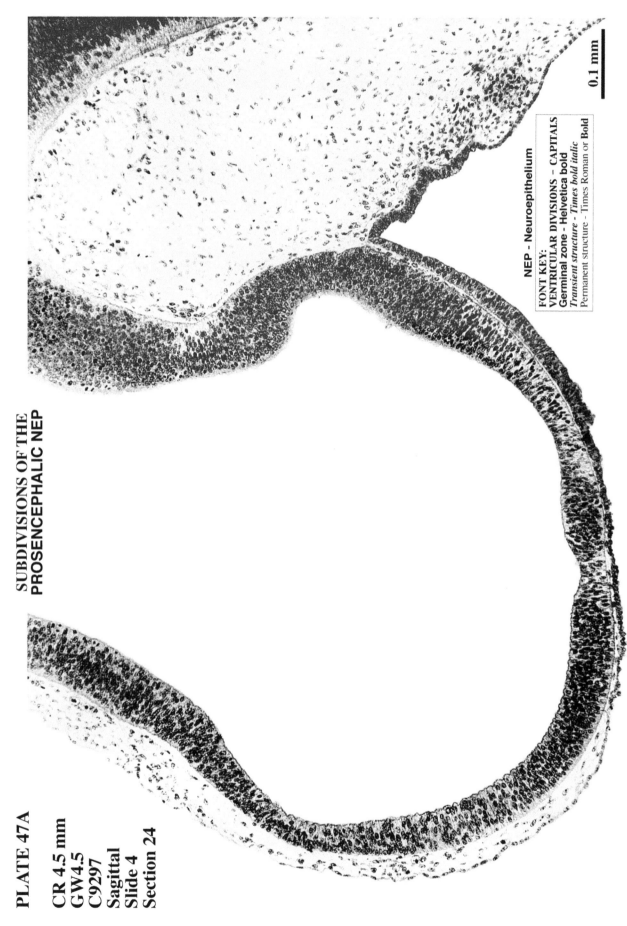

PLATE 47A

CR 4.5 mm
GW4.5
C9297
Sagittal
Slide 4
Section 24

SUBDIVISIONS OF THE
PROSENCEPHALIC NEP

NEP - Neuroepithelium

FONT KEY:
VENTRICULAR DIVISIONS – CAPITALS
Germinal zone - Helvetica bold
Transient structure - Times bold italic
Permanent structure - Times Roman or **Bold**

0.1 mm

See the entire section in Plates 39A and B.

111

PLATE 47B

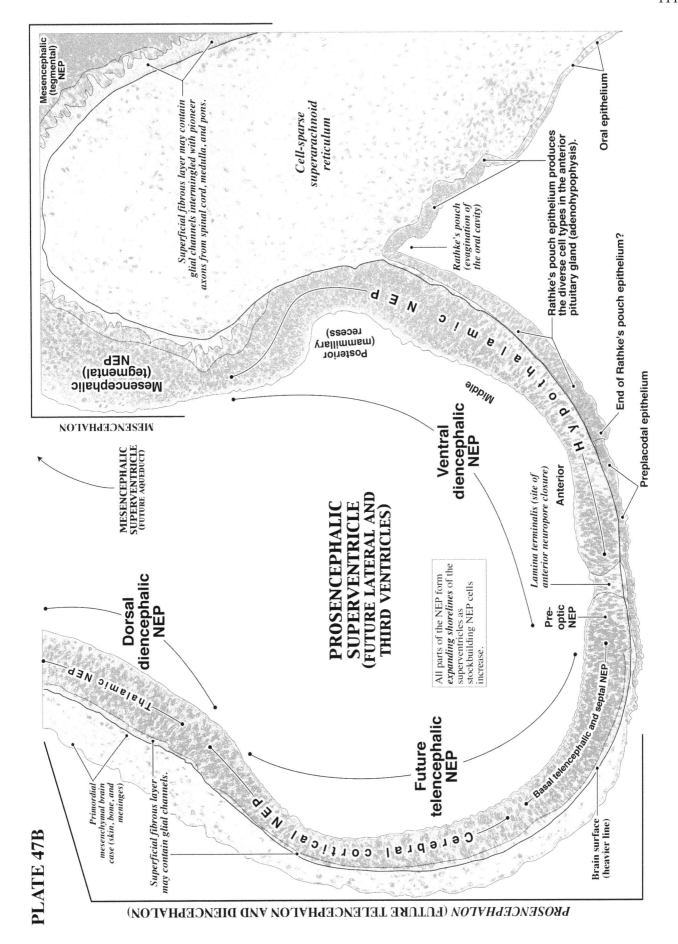

PROSENCEPHALON (FUTURE TELENCEPHALON AND DIENCEPHALON)

MESENCEPHALON

Mesencephalic (tegmental) NEP

Superficial fibrous layer may contain glial channels intermingled with pioneer axons from spinal cord, medulla, and pons.

Cell-sparse superarachnoid reticulum

Oral epithelium

Rathke's pouch epithelium produces the diverse cell types in the anterior pituitary gland (adenohypophysis).

Rathke's pouch (evagination of the oral cavity)

End of Rathke's pouch epithelium

Preplacodal epithelium

Mesencephalic (tegmental) NEP

Posterior (mammillary recess)

H y p o t h a l a m i c   NEP

Middle

Anterior

Lamina terminalis (site of anterior neuropore closure)

MESENCEPHALIC SUPERVENTRICLE (FUTURE AQUEDUCT)

Ventral diencephalic NEP

PROSENCEPHALIC SUPERVENTRICLE (FUTURE LATERAL AND THIRD VENTRICLES)

All parts of the NEP form *expanding shorelines* of the superventricles as stockbuilding NEP cells increase.

Pre-optic NEP

Basal telencephalic and septal NEP

Brain surface (heavier line)

Dorsal diencephalic NEP

T h a l a m i c   NEP

Primordial mesenchymal brain case (skin, bone, and meninges)

Superficial fibrous layer may contain glial channels.

Future telencephalic NEP

C e r e b r a l   c o r t i c a l   NEP

ISTHMUS, CEREBELLUM, AND PONS

PLATE 48A

CR 4.5 mm
GW4.5
C9297
Sagittal
Slide 4, Section 24

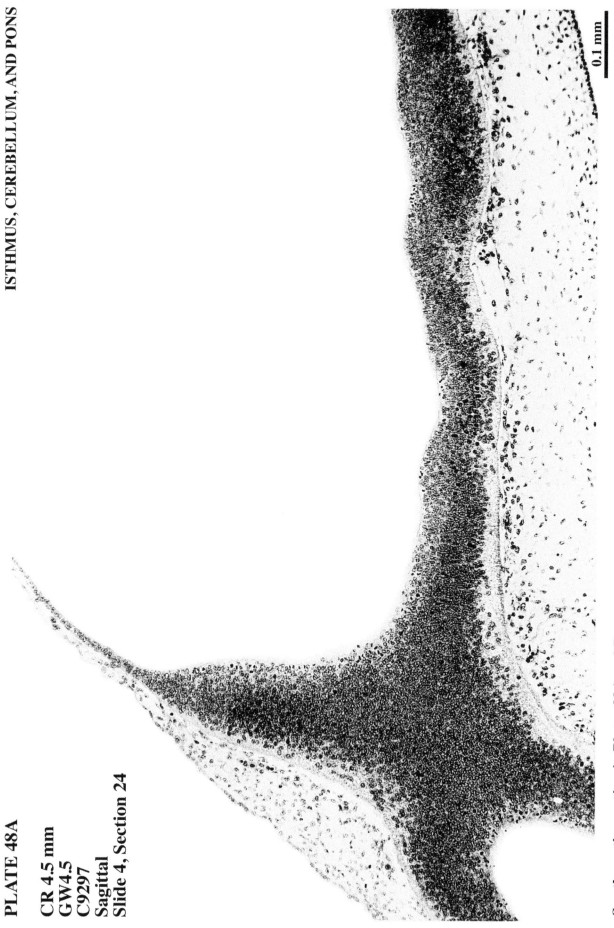

0.1 mm

See the entire section in Plates 39A and B.

113

PLATE 48B

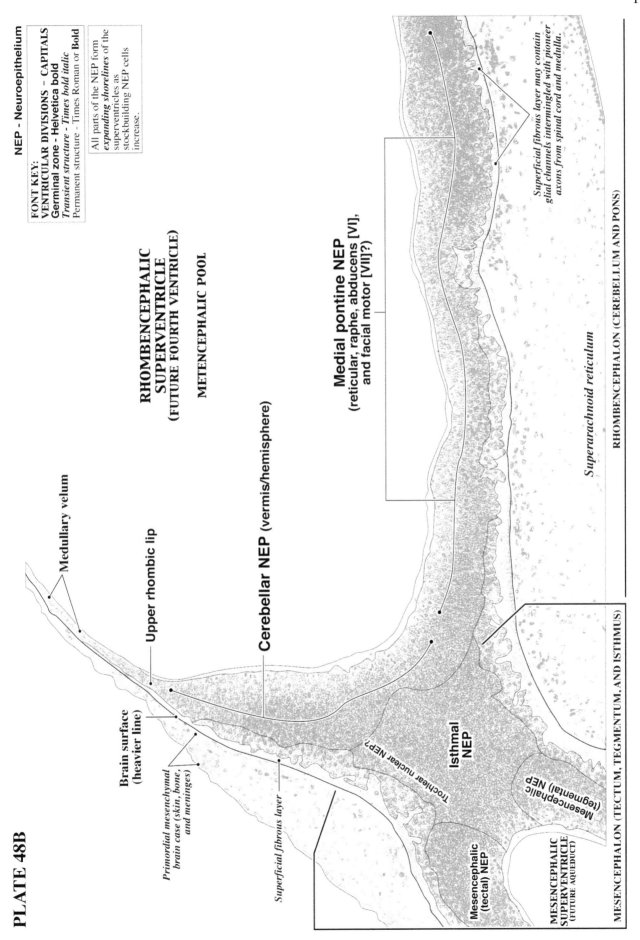

FONT KEY:
VENTRICULAR DIVISIONS – CAPITALS
Germinal zone – **Helvetica bold**
Transient structure – *Times bold italic*
Permanent structure – Times Roman or **Bold**

All parts of the NEP form *expanding shorelines* of the superventricles as stockbuilding NEP cells increase.

NEP – Neuroepithelium

RHOMBENCEPHALIC SUPERVENTRICLE (FUTURE FOURTH VENTRICLE)

METENCEPHALIC POOL

Medial pontine NEP (reticular, raphe, abducens [VI], and facial motor [VII]?)

Superficial fibrous layer may contain glial channels intermingled with pioneer axons from spinal cord and medulla.

Superarachnoid reticulum

RHOMBENCEPHALON (CEREBELLUM AND PONS)

Medullary velum

Upper rhombic lip

Cerebellar NEP (vermis/hemisphere)

Brain surface (heavier line)

Primordial mesenchymal brain case (skin, bone, and meninges)

Superficial fibrous layer

Trochlear nuclear NEP?

Isthmal NEP

Mesencephalic (tegmental) NEP

Mesencephalic (tectal) NEP

MESENCEPHALIC SUPERVENTRICLE (FUTURE AQUEDUCT)

MESENCEPHALON (TECTUM, TEGMENTUM, AND ISTHMUS)

RHOMBOMERES IN PONS AND MEDULLA

PLATE 49A

CR 4.5 mm
GW4.5
C9297
Sagittal
Slide 3, Section 40

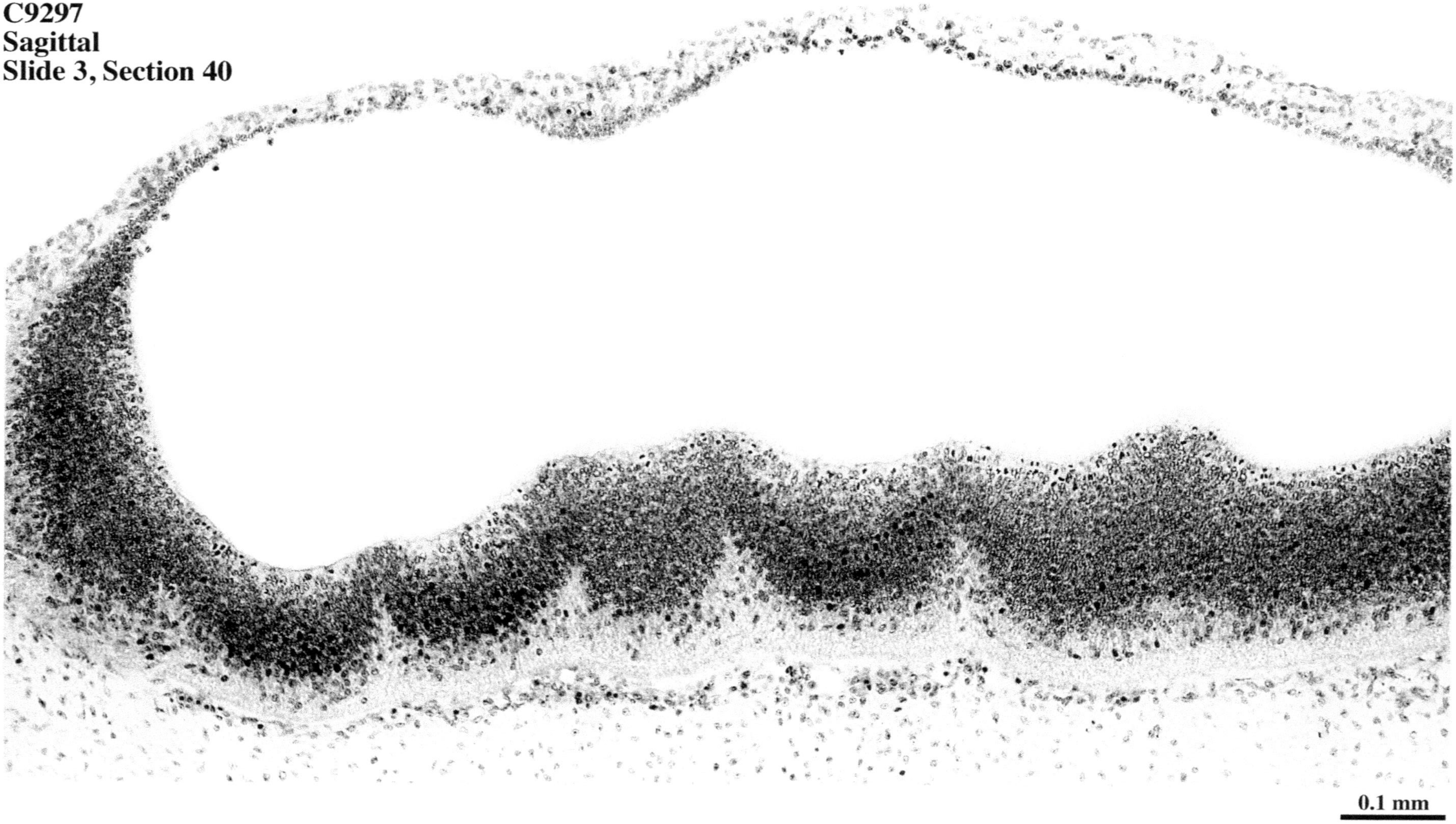

0.1 mm

See the entire section in Plates 42A and B.

PLATE 49B

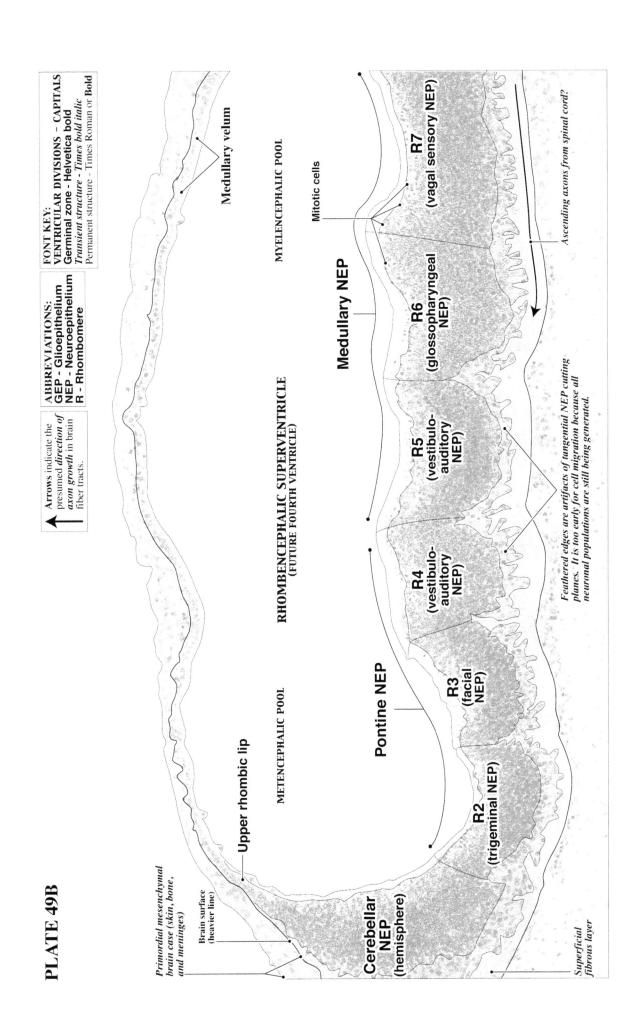

FONT KEY:
VENTRICULAR DIVISIONS – CAPITALS
Germinal zone - Helvetica bold
Transient structure - Times bold italic
Permanent structure - Times Roman or **Bold**

ABBREVIATIONS:
GEP - Glioepithelium
NEP - Neuroepithelium
R - Rhombomere

Arrows indicate the presumed *direction of axon growth* in brain fiber tracts.

Medullary velum

MYELENCEPHALIC POOL

Mitotic cells

R7
(vagal sensory NEP)

Medullary NEP

R6
(glossopharyngeal NEP)

Ascending axons from spinal cord?

RHOMBENCEPHALIC SUPERVENTRICLE
(FUTURE FOURTH VENTRICLE)

R5
(vestibulo-auditory NEP)

R4
(vestibulo-auditory NEP)

Feathered edges are artifacts of tangential NEP cutting planes. It is too early for cell migration because all neuronal populations are still being generated.

Upper rhombic lip

METENCEPHALIC POOL

Pontine NEP

R3
(facial NEP)

R2
(trigeminal NEP)

Primordial mesenchymal brain case (skin, bone, and meninges)

Brain surface (heavier line)

Cerebellar NEP
(hemisphere)

Superficial fibrous layer

RHOMBENCEPHALON AND
SENSORY CRANIAL NERVE ENTRY ZONES

PLATE 50A

CR 4.5 mm, GW 4.5, C9297
Sagittal, Slide 3, Section 24

See the entire section in Plates 44A and B.

0.1 mm

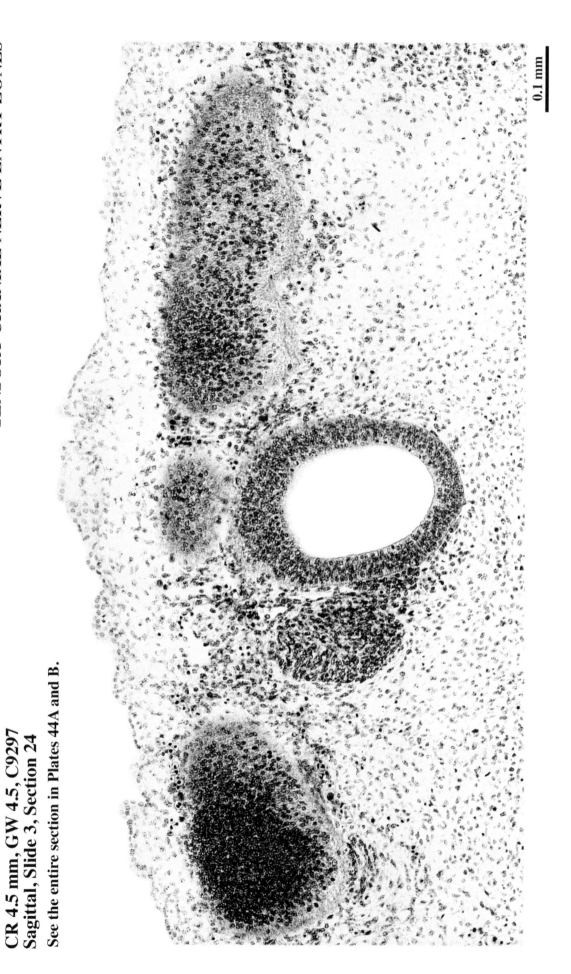

PLATE 50B

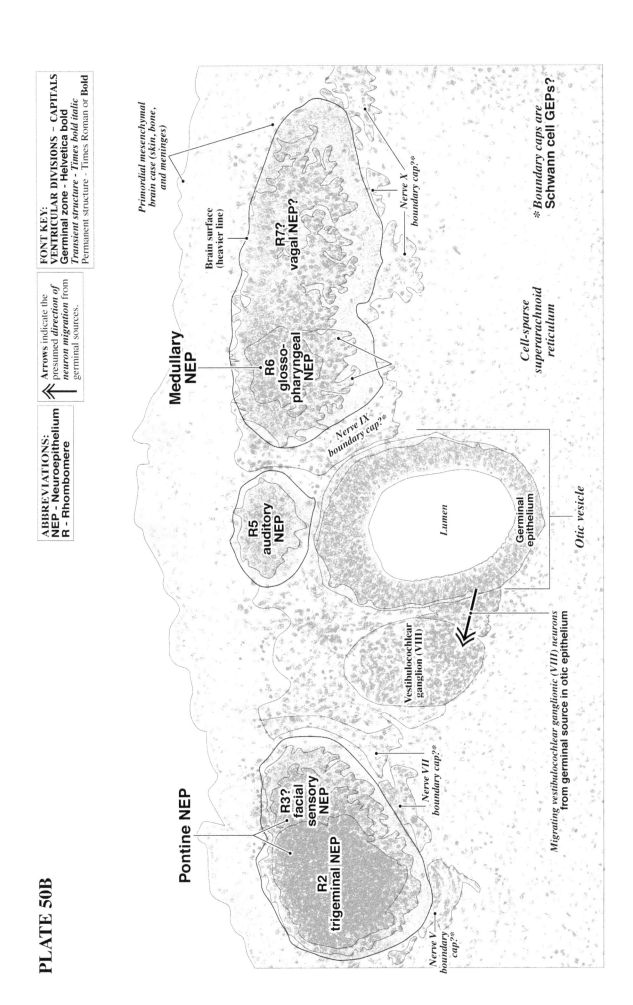

Primordial mesenchymal brain case (skin, bone, and meninges)

Nerve X boundary cap?*

* Boundary caps are Schwann cell GEPs?

Brain surface (heavier line)

R7? vagal NEP?

Medullary NEP

R6 glosso-pharyngeal NEP

Cell-sparse superarachnoid reticulum

Nerve IX boundary cap?*

R5 auditory NEP

Germinal epithelium

Lumen

Otic vesicle

Vestibulocochlear ganglion (VIII)

Pontine NEP

R3? facial sensory NEP

R2 trigeminal NEP

Nerve VII boundary cap?*

Migrating vestibulocochlear ganglionic (VIII) neurons from germinal source in otic epithelium

Nerve V boundary cap?*

118

LOWER MEDULLA AND SPINAL CORD
(MIDLINE RAPHE GLIAL STRUCTURE)

PLATE 51A

CR 4.5 mm
GW4.5, C9297
Sagittal
Slide 4, Section 24

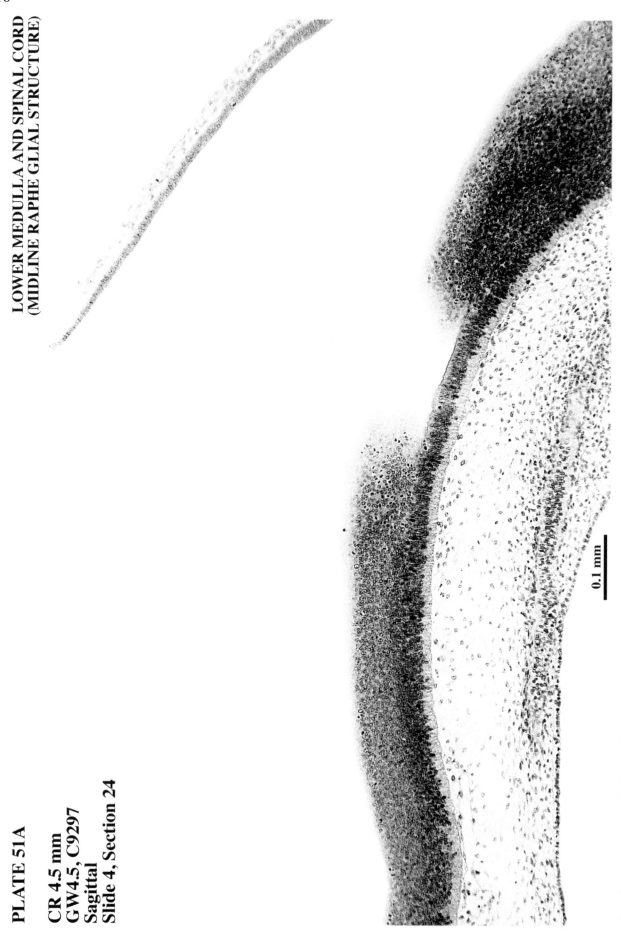

0.1 mm

See the entire section in Plates 39A and B.

PLATE 51B

ABBREVIATIONS:
GEP - Glioepithelium
NEP - Neuroepithelium

FONT KEY:
VENTRICULAR DIVISIONS – CAPITALS
Germinal zone - Helvetica bold
Transient structure - Times bold italic
Permanent structure - Times Roman or Bold

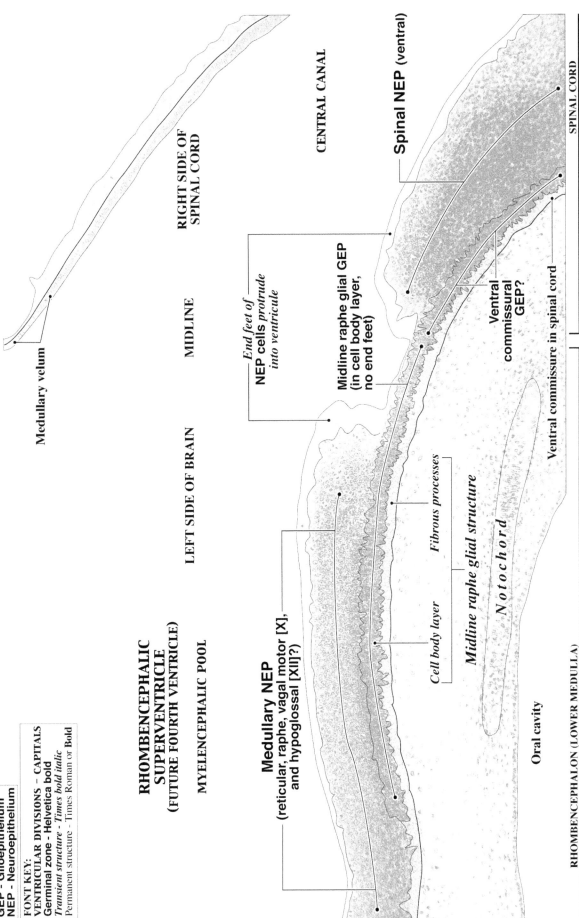

Medullary velum

RIGHT SIDE OF
SPINAL CORD

MIDLINE

CENTRAL CANAL

Spinal NEP (ventral)

End feet of
NEP cells protrude
into ventricule

Midline raphe glial GEP
(in cell body layer,
no end feet)

Ventral
commissural
GEP?

Ventral commissure in spinal cord

SPINAL CORD

LEFT SIDE OF BRAIN

**RHOMBENCEPHALIC
SUPERVENTRICLE**
(FUTURE FOURTH VENTRICLE)

MYELENCEPHALIC POOL

Medullary NEP
(reticular, raphe, vagal motor [X],
and hypoglossal [XII]?)

Fibrous processes

Midline raphe glial structure

Cell body layer

N o t o c h o r d

Oral cavity

RHOMBENCEPHALON (LOWER MEDULLA)

9 781032 183251